AN ILLUSTRATED
INTRODUCTION TO
TOPOLOGY
and
HOMOTOPY

SOLUTIONS MANUAL
FOR PART 1
TOPOLOGY

SASHO KALAJDZIEVSKI

IN COLLABORATION WITH

DEREK KREPSKI
DAMJAN KALAJDZIEVSKI

CRC Press
Taylor & Francis Group
Boca Raton London New York

CRC Press is an imprint of the
Taylor & Francis Group, an **informa** business
A CHAPMAN & HALL BOOK

CRC Press
Taylor & Francis Group
6000 Broken Sound Parkway NW, Suite 300
Boca Raton, FL 33487-2742

International Standard Book Number-13: 978-1-138-55346-0 (Hardback)

Visit the Taylor & Francis Web site at
http://www.taylorandfrancis.com

and the CRC Press Web site at
http://www.crcpress.com

PREFACE AND ACKNOWLEDGEMENT

This solution manual accompanies the first part of the book *An Illustrated Introduction to Topology and Homotopy* by the same author. Except for a small number of exercises in the first few sections, we provide solutions of the (228) odd-numbered problems appearing in the first part of the book (Topology). The primary targets of this manual are the students of topology. This set is not disjoint from the set of instructors of topology courses, who may also find this manual useful as a source of examples, exam problems, etc.

The help of the two collaborators was invaluable to me. However, all typos and errors are mine. Comments related to the book or to the solution manual will be appreciated; please email to sasho@umanitoba.ca. The web page for the book(s) is http://home.cc.umanitoba.ca/~sasho/sk/topology_homotopy.html; corrections are posted there.

I am thankful to Mladen Despic for his help.

SK

Chapter 1: Sets, Numbers and Cardinals

1.1 Sets and Numbers.

Solutions of some exercises

---

2. Given a set X, show that the relation $\subset$ is an order of the set of all subsets of X. For which sets X is this order linear?

Solution. If $A, B \subset X$ are such that $A \subset B$ and $A \neq B$, then there is $b \in B$ such that $b \notin A$. Consequently B is not a subset of A, and hence $\subset$ is antisymmetric. If $A \subset B \subset C$ then obviously $A \subset C$, and so $\subset$ is transitive.

     If X has at least two elements, say a and b, then neither $\{a\} \subset \{b\}$ nor $\{b\} \subset \{a\}$, so the order $\subset$ is not linear. On the other hand if X has at most one element, then the only subsets of X are X and $\varnothing$, and we then readily see that the order $\subset$ is linear.

3. Describe a linear order over (a) the set $\mathbb{N}^2$, and (b) the set $\mathbb{R}^2$.

Solution for (a). Define $(n,m) < (p,q)$ if $n < p$ or ($n = p$ and $m < q$). The parentheses in the preceding sentence are to guarantee there is unique interpretation of the statement that defines $<$. It is left to the reader to prove this relation is antisymmetric and transitive.

4. Show that if $\sim$ is an equivalence relation over a set X, then every two equivalence classes are either disjoint or equal.

Solution. Suppose $[x]$ and $[y]$ are two equivalence classes, and suppose $[x] \cap [y] \neq \varnothing$. Then there is $a \in [x] \cap [y]$. Take any $z \in [x]$. Then $a \sim x \sim z$, and hence $a \sim z$. On the other hand, $a \in [y]$ implies that $y \sim a$. The transitivity of $\sim$ applied to $y \sim a$ and $a \sim z$ yields $y \sim z$. Hence $z \in [y]$. We proved that $[x] \subset [y]$. By the symmetry of the argument, it follows that $[y] \subset [x]$. Hence $[x] = [y]$.

7. Let X be a non-empty set and let $f : X \to Y$ be any mapping. Show that "$u \sim v$ if and only if $f(u) = f(v)$" defines an equivalence relation over X.

Solution. (i) Reflexivity: $u \sim u$ for every u, since $f(u) = f(u)$ for every u. (ii) Symmetry: Suppose $u \sim v$. Then $f(u) = f(v)$, hence $f(v) = f(u)$, hence $v \sim u$. (iii) Transitivity:

Suppose $u \sim v$ and $v \sim w$. Then $f(u) = f(v)$ and $f(v) = f(w)$. Hence $f(u) = f(w)$, and we conclude that $u \sim w$.

1.2 Sets and Cardinal Numbers

Solutions of the odd-numbered exercises

---

1. Let X be an infinite set. Show that for every finite subset A of X, $|X \setminus A| = |X|$. Show that there is a subset B of X such that $|B| = \aleph_0$ and such that $|X \setminus B| = |X|$.

Solution of the first claim. Denote $A = \{a_1, a_2, \ldots, a_n\}$. Use the assumption that X is infinite and induction to construct an infinite countable subset $A_1 = \{a_1, a_2, \ldots, a_n, a_{n+1}, \ldots\}$ of X. The mapping $f(a_k) = a_{k+n}$ defines a bijection from A_1 onto $A_1 \setminus A = \{a_{n+1}, a_{n+2}, \ldots\}$. Then the mapping to $g : X \to X \setminus A$ defined by

$$g(x) = \begin{cases} f(x) & \text{if } x \in A \\ x & \text{if } x \in X \setminus A \end{cases} \quad \text{is a bijection.}$$

3. Let $|A| = |A_1|$, $|B| = |B_1|$, let S be the set of all mappings $A \to B$, and let S_1 be the set of all mappings $A_1 \to B_1$. Show that $|S_1| = |S|$.

Solution. By assumption there exist bijections $\alpha : A \to A_1$ and $\beta : B \to B_1$. Define $\phi : S \to S_1$ as follows: for every $f \in S$, $\phi(f) : A_1 \to B_1$ $\phi(f) = \beta \circ f \circ \alpha^{-1}$. Notice that $\phi(f) : A_1 \to B_1$.

We now check that ϕ is a bijection.

One-to-one: Suppose $\phi(f_1) = \phi(f_2)$. Then $\beta \circ f_1 \circ \alpha^{-1} = \beta \circ f_2 \circ \alpha^{-1}$, so $\beta^{-1} \circ \beta \circ f_1 \circ \alpha^{-1} \circ \alpha = \beta^{-1} \circ \beta \circ f_1 \circ \alpha^{-1} \circ \alpha$, and so $f_1 = f_2$.

Onto: Choose any $g \in S_1$ and let $f = \beta^{-1} \circ g_1 \circ \alpha$. Then

$$\phi(f) = \beta \circ \left(\beta^{-1} \circ g \circ \alpha \right) \circ \alpha^1 = g.$$

5. Prove Proposition 4:

    (a) If J is countable and if each A_j, $j \in J$, is countable, then so is $\bigcup_{j \in J} A_j$.

    (b) If for every $i \in \{1, 2, \ldots, n\}$ the set X_i is countable, then so is the set product $X_1 \times X_2 \times \ldots \times X_n$.

Hint for part (a): Use an argument based on Illustration 1.4 and Proposition 2.

Solution of part (b): Use induction on n. The case when $n = 1$ is trivial. Suppose $Y = X_1 \times X_2 \times \cdots \times X_k$ and X_{k+1} are countable. Hence we can write $Y = \{y_1, y_2, \ldots, y_n, \ldots\}$ and $X_{k+1} = \{x_1, x_2, \ldots, x_n, \ldots\}$. We want to show that $Y \times X_{k+1}$ is countable. The elements of $Y \times X_{k+1}$ are listed (without repetition) in the following two-dimensional array:

$$(y_1, x_1) \quad (y_1, x_2) \quad (y_1, x_3) \quad \cdots$$
$$(y_2, x_1) \quad (y_2, x_2) \quad (y_2, x_3) \quad \cdots$$
$$(y_3, x_1) \quad (y_3, x_2) \quad (y_3, x_3) \quad \cdots$$
$$\vdots \qquad\quad \vdots \qquad\quad \vdots \qquad \ddots$$

Now use the argument as in the caption of Illustration 1.4.

7. Prove that if $|A| = |B|$, then $|\mathcal{P}(A)| = |\mathcal{P}(B)|$.

Solution. Since $|A| = |B|$, there is a bijection $f : A \to B$. Define $\phi : \mathcal{P}(A) \to \mathcal{P}(B)$ by $\phi(X) = f(X)$, for every $X \subset A$. Now check that ϕ is a bijection.

9. Let $|A| = n$ and $|B| = m$ (where n, m are any cardinal numbers). Define $m + n$ to be $|A_1 \cup B_1|$, where A_1 and B_1 are any two disjoint copies of A and B respectively (see the extra problem given below). Show that this operation is well defined (i.e., show it does not depend on the choice of A_1 and B_1). Show that $2 + 3 = 5$.

Solution. Suppose A_2 and B_2 are disjoint copies of A and B respectively. Then there are bijections $f_1 : A \to A_1$, $f_2 : A \to A_2$, $g_1 : B \to B_1$ and $g_2 : B \to B_2$. Define

$$h : A_1 \cup B_1 \to A_2 \cup B_2 \text{ by } h(x) = \begin{cases} f_2 \circ f_1^{-1}(x) & \text{if } x \in A_1 \\ g_2 \circ g_1^{-1}(x) & \text{if } x \in B_1 \end{cases}. \text{ It is now very easy to show that}$$

h is a bijection. Hence $|A_1 \cup B_1| = |A_2 \cup B_2|$ and thus $m + n$ does not depend on the choice of the copies A_1 and B_1. Now, to show that $2 + 3 = 5$ it suffices to take any two disjoint copies A_1 and B_1 of $A = \{1,2\}$ and $B = \{1,2,3\}$ respectively, and then simply count the elements of $A_1 \cup B_1$.

11.   (a) Prove that if $n \geq m$, then $2^n \geq 2^m$.
       (b) Prove that if $2^n \geq \aleph_0$, then $2^n \geq 2^{\aleph_0}$.

Solution.

   (a). Choose any set A with $|A| = n$, and any set B such that $|B| = m$. Hence $|\mathcal{P}(A)| = 2^n$ and $|\mathcal{P}(B)| = 2^m$. Since we have assumed that $|A| \geq |B|$ it follows from Proposition 2 that there is an onto mapping $g : A \to B$. Define $\phi : \mathcal{P}(A) \to \mathcal{P}(B)$ by

$\phi(\{a_j : j \in J\}) = \{g(a_j) : j \in J\}$, for every subset $\{a_j : j \in J\}$ of A. Since g is onto, it follows that ϕ is also onto. The conclusion now follows from Proposition 2.

(b). Suppose $2^n \geq \aleph_0$. If n is finite, then so is 2^n. This would then contradict our assumption here. Hence n is infinite, and thus $n \geq \aleph_0$. It follows from part (a) that $2^n \geq 2^{\aleph_0}$.

Extras

1. Given a set A prove that there exists a set B such that $|A| = |B|$ and $A \cap B = \varnothing$.

1.3 Axiom of Choice and Equivalent Statements

Solutions of some exercises

1. Prove that the Axiom of Choice and the Axiom for Products are equivalent.

Solution. Suppose the Axiom of Choice is true. We want to show that if $A_i \neq \varnothing$ for every $i \in I$, then $\prod_{i \in I} A_i \neq \varnothing$. By the Axiom of Choice there is a mapping $f : I \to \bigcup_{i \in I} A_i$ such that $f(i) \in A_i$ for every $i \in I$. Then $(f(i))_{i \in I}$ is an element of the product $\prod_{i \in I} A_i$, and so $\prod_{i \in I} A_i \neq \varnothing$. Conversely, let $\{A_i : i \in I\}$ be a family of pairwise disjoint sets, and suppose $\prod_{i \in I} A_i \neq \varnothing$. Then there is some $(a_i)_{i \in I}$ in $\prod_{i \in I} A_i$. Define $f : I \to \bigcup_{i \in I} A_i$ by $f(i) = a_i$. This mapping is such that $f(i) \in A_i$ for every $i \in I$.

4. Prove that if X is linearly ordered by $\leq$, and if every countable subset of X is well ordered by $\leq$, then X is well ordered by $\leq$. [Hint: start by assuming that X is not well ordered, and construct a countable linearly ordered set without the least element.]

Solution. Following the hint, suppose the hypotheses of this problem are fulfilled and suppose that X is not well ordered. This means that there is some $A \subset X$ such that A does not have the least element. Choose any $a_1 \in A$. If there is an $a \in A$ that is not comparable to a_1, then $\{a, a_1\}$ is not well ordered, contradicting the assumption that every countable subset of X is well ordered. So, a_1 is comparable to any other element of A. By assumption, a_1 could not be the least element of A. Hence there is $a_2 \in A$ such that $a_2 < a_1$. Repeat this argument. This procedure cannot end after n many steps, since then a_n would be the least element of A. We got an infinite sequence $a_1 > a_2 > ... > a_n > ...$ of countably many elements of X that does not have the least element. This contradicts our assumptions. Hence X must be well ordered.

5. Prove that for every order R of any set A there exists a linear order Q over A such that $R \subset Q$. [Hint: 1. For every chain $R_1 \subset R_2 \subset \cdots$ of orders over A, $\bigcup_{i=1}^{\infty} R_i$ is a partial order containing all of $R_i - s$. 2. If M is a maximal order, then M must be linear. Use Zorn's lemma.]

Solution: The statement in the part 1 of the hint is evident. The countability suggested by the hint is irrelevant: the analogue statement is true for any chain of orders. By Zorn's lemma, there is a maximal order M of A, such that $R \subset M$. If M is not linear, then there are $a, b \in A$ that are not comparable. We now prove that the existence of such $a, b \in A$ implies that M can be extended to a linear order, contradicting the maximality of M. We fist extend M by setting $a < b$. If this extension is transitive, then it is also an order, and we immediately get a contradiction (to the maximality of M). Suppose it is not transitive. Then there are $x, y \in A$ such that $x < a$ and $b < y$ (in M), yet x is not less than y (we take strict inequalities since the case when $x = a$ or $y = b$ is evident). Hence either $y < x$ in M or x and y are not comparable. If $y < x$ then $b < y < x < a$ in M, and so a and b are comparable, contradicting our assumption. Hence or x and y are not comparable (in M). In this case we simply M by setting $x < y$ for every $x, y \in A$ for which there are a and b as above. This construction gives a linear order of A that extends M, contradicting the maximality of M.

Chapter 2: Metric Spaces: Definition, Examples and Basics

2.1 Metric Spaces: Definition and Examples

Solutions of some exercises

2. Prove that the Euclidean metric space and the city metric over $\mathbb{R}^2$ are equivalent.

Solution. Let U be a non-empty subset of $\mathbb{R}^2$ that is open with respect to the Euclidean metric d, and denote the city metric by d_{cm}. We show that U is open with respect to the city metric. It suffices to show that for every point $\mathbf{x} = (x_1, x_2) \in U$ there is a city-metric ball $B_{cm}(\mathbf{x}, r)$ such that $B_{cm}(\mathbf{x}, r) \subset U$. By assumption, there is a Euclidean ball $B(\mathbf{x}, s)$ such that $B(\mathbf{x}, s) \subset U$. Since it is evident that for every two points $\mathbf{u}, \mathbf{v} \in \mathbb{R}^2$, $d(\mathbf{u}, \mathbf{v}) \leq d_{cm}(\mathbf{u}, \mathbf{v})$, it follows immediately that $B_{cm}(\mathbf{x}, s) \subset B(\mathbf{x}, s)$. Hence $B_{cm}(\mathbf{x}, s) \subset U$, and thus we proved that U is open with respect to the city metric.

Conversely, suppose U is open with respect to d_{cm}. Hence, for every $\mathbf{x} = (x_1, x_2) \in U$ there is a city-metric ball $B_{cm}(\mathbf{x}, r)$ such that $B_{cm}(\mathbf{x}, r) \subset U$. Choose $s = \dfrac{r}{2}$, and consider the Euclidean ball $B(\mathbf{x}, s)$. We want to prove that $B(\mathbf{x}, s) \subset U$, i.e., that for every $\mathbf{y} = (y_1, y_2) \in B(\mathbf{x}, s)$, $\mathbf{y}$ is also in U. So, assume $\mathbf{y} = (y_1, y_2) \in B(\mathbf{x}, s)$. This means that $(x_1 - y_1)^2 + (x_2 - y_2)^2 < s^2 = \dfrac{r^2}{4}$, i.e.,

$\left(|x_1 - y_1| \right)^2 + \left(|x_2 - y_2| \right)^2 < \dfrac{r^2}{4}$. Since $2|x_1 - y_1||x_2 - y_2| \leq \left(|x_1 - y_1| \right)^2 + \left(|x_2 - y_2| \right)^2$, it follows that

$2|x_1 - y_1||x_2 - y_2| < \dfrac{r^2}{4}$. Hence $\left(|x_1 - y_1| \right)^2 + 2|x_1 - y_1||x_2 - y_2| + \left(|x_2 - y_2| \right)^2 < \dfrac{r^2}{4} + \dfrac{r^2}{4} = \dfrac{r^2}{2}$, and

thereby $\left(|x_1 - y_1| + |x_2 - y_2| \right)^2 < \dfrac{r^2}{2}$. Take square-root on both sides to get $|x_1 - y_1| + |x_2 - y_2| < \dfrac{r}{\sqrt{2}}$.

This last inequality implies that $d_{cm}(\mathbf{x}, \mathbf{y}) < \dfrac{r}{\sqrt{2}}$. Hence $d_{cm}(\mathbf{x}, \mathbf{y}) < r$ and consequently $\mathbf{y} \in B_{cm}(\mathbf{x}, r) \subset U$. We proved that $y \in U$, so that $B(\mathbf{x}, s) \subset U$. With this we established U is open with respect to the Euclidean metric, and finished the proof that d and d_{cm} are equivalent.

5. Consider $\mathbb{R}^2$ with the topology induced by the post office metric with respect to a fixed point p.

(a) Show that if $p \notin A \subset \mathbb{R}^2$ then A is open.

(b) Show that if $p \in A \subset \mathbb{R}^2$ then A is open if and only if there is a Euclidean ball $B(p,r)$ (that is, a ball with respect to the Euclidean metric) that is contained in A.

Solution.
(a). Since every set is a union of singletons, it suffices to show that every singleton $\{x\}$, $x \neq p$, is open. That this is true follows from the observation that the open ball around x and of radius $\frac{1}{2}d(x, p)$ contains only x.

(b). Suppose A is open in the post office metric. Then there is a post-office metric ball $B_{po}(p,r)$ that is a subset of A. Since the Euclidean balls and the post office metric balls around p coincide, it follows that $B(p,r) \subset A$. Conversely, suppose $B(p,r) \subset A$ for some r. Since $B(p,r) = B_{po}(p,r)$, it follows $B_{po}(p,r) \subset A$. That there are post-office metric balls around the other points of the set A was established in part (a). Hence A is open in the post office metric space.

6. [Note the typo in the book: min is replaced by max here. Also, take a look at the extra problem below.] Suppose d_1 and d_2 are two metrics over a set X. Show that d_3 defined by $d_3(x,y) = \max\{d_1(x,y), d_2(x,y)\}$ is also a metric over X.

Solution. Going over the three conditions of the definition of a metric:

(i)   It is obvious that $d_3(x,y) \geq 0$ all the time. Further, $d_3(x,x) = \max\{d_1(x,x), d_2(x,x)\} = 0$; conversely, if $d_3(x,y) = 0$, then $\max\{d_1(x,y), d_2(x,y)) = 0$, hence $d_i(x,y) = 0$ for some $i \in \{1,2\}$, and this in turn implies that $x = y$.

(ii) $d_3(x,y) = \max\{d_1(x,y), d_2(x,y)) = \max\{d_1(y,x), d_2(y,x)) = d_3(y,x)$.

(iii) $d_3(x,y) = \max\{d_1(x,y), d_2(x,y)) \leq \max\{d_1(x,z) + d_1(z,y), d_2(x,z) + d_2(z,y)\} \leq$
$\leq \max\{d_1(x,z), d_2(x,z)\} + \max\{d_1(z,y), d_2(z,y)\} = d_3(x,z) + d_3(z,y)$.

8. Let F be a closed subset of a metric space (X,d). For every point $x \in X$, define $d(x,F) = \inf\{d(x,y) : y \in F\}$. Show that $x \in X \setminus F$ if and only if $d(x,F) > 0$.

Solution. Suppose x is out of F. Then x is in F^c, which is an open set. Hence, there is a ball $B(x,r)$ completely within F^c. Hence there is no point in F closer than r units to x, So, $d(x,F) \geq r > 0$. The converse is obvious.

10. Show that if F and G are two disjoint closed sets in a metric space X, then there are two disjoint open sets U and V such that $F \subset U$ and $G \subset V$. [Hint: Exercise 8]

Solution. We use the distance between a point and a closed set as defined in Exercise 8. For every point $x \in F$, choose $r_x = \frac{1}{4} d(x, G)$. We note in passing that any number $< \frac{1}{2}$ would do in place of $\frac{1}{4}$. Then $F \subset U = \bigcup_{x \in F} B(x, r_x)$. Symmetrically, for every $y \in G$ choose $s_y = \frac{1}{4} d(y, F)$; then $G \subset V = \bigcup_{y \in G} B(y, s_y)$. The open sets U and V are disjoint, for if there is $z \in U \cap V$, then $z \in B(x, r_x) \cap B(y, s_y)$, for some $x \in F$ and $y \in G$. We find

$d(x, y) \le d(x, z) + d(z, y) < r_x + s_y = \frac{1}{4} d(x, G) + \frac{1}{4} d(y, F)$. By symmetry we may suppose

that $d(x, G) \le d(y, F)$. Then $d(x, y) \le \frac{1}{4} d(x, G) + \frac{1}{4} d(y, F) \le 2 \frac{1}{4} d(y, F) = \frac{1}{2} d(y, F)$,

which is obviously impossible. Hence $U \cap V = \varnothing$.

11. Consider the set $\mathcal{R}$ of all open rectangles in $\mathbb{R}^2$ above the x-axis, bounded by edges of slopes 1 and -1, with one corner at the x-axis, and such that every point $(x, 0)$ on the x-axis is a corner of exactly one member R_x of $\mathcal{R}$ (see Illustration 2.10). Show that there is a rational number q and an irrational number z such that $R_q \cap R_z \ne \varnothing$.

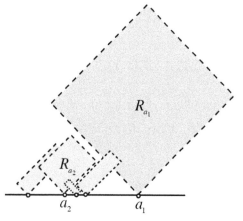

Illustration 2.10. We show a few rectangles from the set $\mathcal{R}$.

Solution. Assume there is a set $\mathcal{R}$ as above such that $R_q \cap R_z = \varnothing$ for every rational number q and irrational number z. Start with a rational number q_1 and the corresponding rectangle R_{q_1}. Choose an irrational number x_1 to the left of q_1, such that

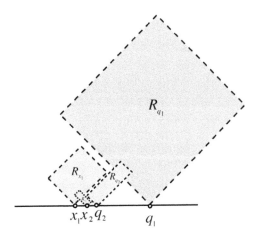

Figure 1. The first steps in the construction of the sequences $\{q_i\}$ and $\{x_i\}$

$q_1 - x_1 < \dfrac{1}{2}$ and such that x_1 belongs to the orthogonal projection of R_{q_1} on the x-axis (see Figure 1). By assumption we have that $R_{q_1} \cap R_{x_1} = \varnothing$. Next, choose a rational number q_2 to the right of x_1, in the orthogonal projection of R_{x_1} on the x-axis, and such that $q_2 - x_1 < \dfrac{1}{4}$. Our assumption implies that $R_{q_2} \cap R_{x_1} = \varnothing$. We iterate this procedure countable many times (see Figure 10) to get two sequences: the decreasing sequence $\{q_i\}$, and the increasing sequence $\{x_i\}$ It follows from our construction that these two sequences converge to a single point w.

Suppose w is rational. Since each q_i is in the orthogonal projection of each R_{x_i}, it follows that w is in the orthogonal projection of each $\overline{R}_{x_i}$. As a consequence, the line of slope -1 passing through w intersects each of R_{x_i}. Now choose any x_i such that R_{x_i} is closer to w than the length of the side of R_w of slope -1. Then $R_{x_i} \cap R_w \neq \varnothing$, and we get a contradiction.

The case when w is irrational is symmetric.

Extras.

1. Suppose d_1 and d_2 are two metrics over a set X. Show that d_3 defined by $d_3(x,y) = \max\{d_1(x,y), d_2(x,y)\}$ need not be a metric over X.

2.2 Metric Spaces: Basics

Solutions of some exercises

---

2. Let (a_n) be a non-decreasing sequence in $\mathbb{R}$ that is bounded from above. Show that (a_n) converges.

Solution. By the Least Upper Bound Property, since the set $\{a_n : n \in \mathbb{N}\}$ is bounded it has an upper bound a. If $a_k = a$ for some k, it follows from our assumptions that $a_n = a$ for every $n \geq k$, and so (a_n) converges to a. Otherwise $a_k < a$ for every k. If a is not the limit of the sequence then there is $\epsilon > 0$ such that the interval $(a - \epsilon, a + \epsilon)$ does not contain any members of the sequence (a_n). That would make $a - \epsilon$ an upper bound for (a_n) that is smaller than a. This is a contradiction, and hence (a_n) converges to a.

3. Prove Proposition 1: Let (A,d) be a metric subspace of (X,d). For every $V \subset A$, V is open in A if and only if there is an open subset U of X such that $V = U \cap A$.

Solution. Suppose V is open in A. Then $V = \bigcup_{i \in I} B_i$, where each B_i is an open ball $B_A(x, r_i) = \{y \in A : d_A(x,y) < r_i\}$ in the metric subspace A. The subspace metric d_A is the restriction of the metric d over A. Hence $B_A(x, r_i) = A \cap \{y \in X : d(x,y) < r_i\} = A \cap B(x, r_i)$, where $B(x, r_i)$ is a ball in X. Backing up, we see that $V = \bigcup_{i \in I} B_i = \bigcup_{i \in I}(A \cap B(x, r_i))$, which, by de Morgan laws, is the same as $A \cap \bigcup_{i \in I} B(x, r_i)$, We can now set $U = \bigcup_{i \in I} B(x, r_i)$. For the converse just follow the above argument in the opposite direction.

5. Show that a subset A of a metric space X is closed if and only if every sequence (a_n) of elements in A that converges in X also converges in A.

Solution. Suppose A is closed, and let (a_n) of elements in A that converges to x in X. If $x \notin A$, then, since the complement of A is open, there is a ball around it completely within that complement. So, that ball will avoid all (a_n), contradicting the assumption that (a_n) converges to x. So, $x \in A$. Conversely, suppose every sequence (a_n) of elements of A converging in X, converges in A. We show that A is closed, i.e., that A^c is open. Take $a \in A^c$, and consider the balls $B\left(a, \dfrac{1}{n}\right)$. If A^c is not open, then each of them

intersects A. Choose an element a_n in each $B\left(a, \dfrac{1}{n}\right) \cap A$. Then (a_n) is a sequence in A

converging to $a \notin A$, and we have a contradiction.

8. Show that if (x_n) is a Cauchy sequence that has a convergent subsequence, then (x_n) is also convergent.

Solution. Suppose $\left(x_{m_i}\right)$ is a convergent subsequence of the Cauchy sequence (x_n), and suppose

it converges to a. Choose any $\epsilon > 0$, and let N be such that if $n, m > N$, then $\left|x_n - x_m\right| < \dfrac{\epsilon}{2}$. This

can be done since (x_n) is a Cauchy sequence. Since the subsequence $\left(x_{m_i}\right)$ converges to a, there

is M such that for every $m_i > M$, $\left|x_{m_k} - a\right| < \dfrac{\epsilon}{2}$. Let L be $\max\{M, N\}$. Then for every $n > L$ we

have $\left|x_n - a\right| = \left|x_n - x_{m_i} + x_{m_i} - a\right| \le \left|x_n - x_{m_i}\right| + \left|x_{m_i} - a\right| < \dfrac{\epsilon}{2} + \dfrac{\epsilon}{2} = \epsilon$, which confirms that (x_n) also

converges to a.

9. Let (X_i, d_i), $i = 1, 2, \ldots, n$, be a set of metric spaces and denote $X = \displaystyle\prod_{i=1}^{n} X_i$, their product metric

space. Show that every projection p_i and every coordinate mapping $c_i^{\mathbf{a}}$ is continuous.

Solution. We use Proposition 3. Let $(\mathbf{x}_m)$ be a sequence of elements of X converging to

$\mathbf{y} = (y_1, y_2, \ldots, y_n) \in X$, and let $p_k : \displaystyle\prod_{i=1}^{n} X_i \to X_k$ be the projection. If $\mathbf{x}_m = (x_{m_1}, x_{m_2}, \ldots, x_{m_n})$, then

the image of the sequence $(\mathbf{x}_m)$ under p_k is the sequence $\left(x_{k_j}\right)_{j=1}^{\infty}$. That this sequence converges

to y_k follows from the assumption that $(\mathbf{x}_m)$ converges to $\mathbf{y} = (y_1, y_2, \ldots, y_n)$. (See exercise 14 below.)

Proving that the coordinate mappings are continuous can be done via a similar argument.

10. Let (X_i, d_i), $i = 1, 2, \ldots, n$, be metric spaces, and consider the set product $X = \displaystyle\prod_{i=1}^{n} X_i$. Define

$d^{\#}((x_1, x_2, \ldots, x_n), (y_1, y_2, \ldots, y_n)) = d_1(x_1, y_1) + d_2(x_2, y_2) + \ldots + d_n(x_n, y_n)$, for every

$(x_1, x_2, \ldots, x_n), (y_1, y_2, \ldots, y_n) \in X$. Show that $d^{\#}$ is a metric on the set X which is equivalent to the product metric over X.

Solution. (Observe that this problem generalizes Exercise 2 in Section 2.1, since $d^{\#}$ generalizes the city metric.)

We show that $d^{\#}$ is equivalent to the product metric over X. That it is a metric for $n = 2$ is proven in Example 4, Section 2.1.

First we prove that every set open with respect to the product metric d_p is open with respect to $d^{\#}$.

Since $\left(d_1(x_1, y_1) + d_2(x_2, y_2) + \cdots + d_n(x_n, y_n)\right)^2 \geq d_1(x_1, y_1)^2 + d_2(x_2, y_2)^2 + \cdots + d_n(x_n, y_n)^2$,

We have $d_1(x_1, y_1) + d_2(x_2, y_2) + \cdots + d_n(x_n, y_n) \geq \left(d_1(x_1, y_1)^2 + d_2(x_2, y_2)^2 + \cdots + d_n(x_n, y_n)^2\right)^{\frac{1}{2}}$, and

so $d_p(\mathbf{x}, \mathbf{y}) \leq d^{\#}(\mathbf{x}, \mathbf{y})$ all the time (where d_p is the product metric). Denote balls with respect to $d^{\#}$ by $B^{\#}$. It follows from $d_p(\mathbf{x}, \mathbf{y}) \leq d^{\#}(\mathbf{x}, \mathbf{y})$ that $B(\mathbf{x}, r) \subset B^{\#}(\mathbf{x}, r)$ all the time, and hence every open set with respect to $d^{\#}$ is open with respect to d_p. (See Exercise 1 in Extras.)

Now we prove the converse: that every set open with respect to d_p is open with respect to $d^{\#}$. It suffices to show that every ball $B(\mathbf{x}, r)$ contains some ball $B^{\#}(\mathbf{x}, s)$. First we notice that

$$\left(d_1(x_1, y_1) + d_2(x_2, y_2) + \cdots + d_n(x_n, y_n)\right)^2 = \sum_{i=1}^{n} d_i(x_i, y_i)^2 + 2 \sum_{\text{all } i \neq j} d_i(x_i, y_i) d_j(x_j, y_j) \leq$$

$$\leq \sum_{i=1}^{n} d_i(x_i, y_i)^2 + m\, d_{\max}(x_{i_0}, y_{i_0})^2, \text{where } d_{\max}(x_{i_0}, y_{i_0}) = \max\{d_i(x_i, y_i) : i = 1, 2, \ldots, n\}, \text{ and where } m$$

is twice the number of all combinations of i and j, $i, j \in \{1, 2, \ldots, n\}$. The next step is evident:

$$\sum_{i=1}^{n} d_i(x_i, y_i)^2 + m\, d_{\max}(x_{i_0}, y_{i_0})^2 \leq (m+1) \sum_{i=1}^{n} d_i(x_i, y_i)^2. \text{ Hence}$$

$$\left(d_1(x_1, y_1) + d_2(x_2, y_2) + \cdots + d_n(x_n, y_n)\right)^2 \leq (m+1) \sum_{i=1}^{n} d_i(x_i, y_i)^2, \text{ and consequently}$$

$$\left(d_1(x_1, y_1) + d_2(x_2, y_2) + \cdots + d_n(x_n, y_n)\right) \leq \sqrt{m+1} \left(\sum_{i=1}^{n} d_i(x_i, y_i)^2\right)^{\frac{1}{2}}, \text{ that is,}$$

$\dfrac{1}{\sqrt{m+1}} d^{\#}(\mathbf{x}, \mathbf{y}) \leq d_p(\mathbf{x}, \mathbf{y})$ all the time. It follows that for every ball $B(\mathbf{x}, r)$, we have

$B^{\#}\left(\mathbf{x}, \dfrac{r}{\sqrt{m+1}}\right) \subset B(\mathbf{x}, r)$, and we have established what we were after.

14. Let (X_1, d_1) and (X_2, d_2) be metric spaces and let $\{(x_i, y_i)\}$ be a sequence in the product metric space $X_1 \times X_2$. Show that $\lim_{i \to \infty}(x_i, y_i) = (a, b)$ in $X_1 \times X_2$ if and only if $\lim_{i \to \infty} x_i = a$ in X_1 and $\lim_{i \to \infty} y_i = b$ in X_2.

Solution. Suppose $\lim_{i \to \infty}(x_i, y_i) = (a, b)$. Then, given any $r > 0$, for $i > N$, all (x_i, y_i) are in a ball of radius r and centered at (a, b). Hence, from that point on, we have $\sqrt{d_1(x_i, a)^2 + d_2(y_i, b)^2} < r$. But that obviously implies that, for $i > N$, we have $d_1(x_i, a) < r$ and $d_2(y_i, a) < r$, proving that $\lim_{i \to \infty} x_i = a$ in X_1 and $\lim_{i \to \infty} y_i = b$ in X_2.

Conversely, suppose $\lim_{i \to \infty} x_i = a$ in X_1 and $\lim_{i \to \infty} y_i = b$ in X_2. Start with any $r > 0$. Then there exists N_1 such that $d_1(x_i, a) < \dfrac{r}{\sqrt{2}}$ for all $i > N_1$, and there exists N_2 such that $d_2(y_i, a) < \dfrac{r}{\sqrt{2}}$ for all $i > N_2$. Denote $N = \max\{N_1, N_2\}$. Then for every $i > N$, $d_1(x_i, a) < \dfrac{r}{\sqrt{2}}$ and $d_2(y_i, a) < \dfrac{r}{\sqrt{2}}$. It follows that, for every $i > N$, we have $d_1(x_i, a)^2 + d_2(y_i, b)^2 < \dfrac{r^2}{2} + \dfrac{r^2}{2} = r^2$. Hence for every $i > N$, we have $\sqrt{d_1(x_i, a)^2 + d_2(y_i, b)^2} < r$. We conclude that $\lim_{i \to \infty}(x_i, y_i) = (a, b)$.

16. Let $f : [0,1] \to [0,1]$ be a continuous function. Show that $g : [0,1] \to [0,1]$ defined by $g(t) = \sup\{f(x) : x \in [0,t]\}$ is continuous and non-decreasing. Draw a sketch comparing the graphs of f and g if $f(x) = -\sin(2\pi x)$.

Solution. That g is non-decreasing is obvious. For continuity take a sequence $\{x_n\}$ converging to y. We need to show that $\{a_n\} = \{\sup\{f(x) : x \in [0, x_n]\}\}$ converges to $a = \sup\{f(x) : x \in [0, y]\}$. Start with $\epsilon > 0$. We are searching for N, such that for every $n > N$, $|a_n - a| < \epsilon$.

We argue differently depending on whether (i) $g(y) = f(y)$ or (ii) $f(y) < g(y)$.

Assume (i). Let $\epsilon > 0$ be given and suppose $N > 0$ is such that $| f(x_n) - f(y)| < \epsilon$ (i.e., $f(x_n) \in (f(y) - \epsilon, f(y) + \epsilon)$) whenever $n > N$.

For such n, if $x_n < y$, then we have $g(x_n) \le g(y)$ since g is non-decreasing, and $g(y) - \epsilon = f(y) - \epsilon < f(x_n) \le g(x_n)$ since $f \le g$ everywhere. That is, $g(x_n) \in (g(y) - \epsilon, g(y)]$.

Similarly, if $n > N$ and $y < x_n$, we have $g(y) \leq g(x_n)$. Moreover, since we assume (i), $g(x_n)$ equals the maximum value obtained by f on $[y, x_n]$, which must be within ϵ of $f(y)$ (by virtue of being a value attained by f). Hence $g(x_n) \in (g(y), g(y) + \epsilon)$.

Summarizing, if (i) and $n > N$, then we have $g(x_n) \in (g(y), g(y) + \epsilon)$.

Assume (ii). Then by continuity of f, there exists a small neighborhood of y such that $f(t) < g(y)$ for all t in that neighborhood. Therefore, $g(t)$ is constant on that neighborhood so that $g(x_n)$ is eventually constant and hence convergent.

18. Let (X, d_1) and (X, d_2) be two metric spaces (over the same set X), let $X = A \cup B$ and let $A \cap B = C \neq \varnothing$. Assume also that $d_1(c_1, c_2) = d_2(c_1, c_2)$ for every $c_1, c_2 \in C$.

Define $d : X \times X \to \mathbb{R}$ as follows: $d(x, y) = \begin{cases} d_1(x, y) & \text{if } x, y \in A \\ d_2(x, y) & \text{if } x, y \in B \\ \inf\{d_1(x, z) + d_2(y, z) : z \in C\} & \text{otherwise} \end{cases}$

Show that d is a metric.

Solution. First of all, for the infimum to exist, the set of the numbers $\{d_1(x, z) + d_2(y, z) : z \in C\}$ has to be bounded from below, which is obvious (bounded by 0). So, that is fine. The only problematic property is the triangle inequality: $d(x, y) + d(y, u) \leq d(x, u)$. This is fine if all of x, y, u are in A or if all of them are in B. Suppose $x \in A$ and $y, u \in B$. Then

$d(x, y) + d(y, u) = \inf\{d_1(x, z) + d_2(y, z) : z \in C\} + d_2(y, u)$ while

$d(x, u) = \inf\{d_1(x, z) + d_2(u, z) : z \in C\}$. We have

$\inf\{d_1(x, z) + d_2(y, z) : z \in C\} + d_2(y, u) = \inf\{d_1(x, z) + d_2(y, z) + d_2(y, u) : z \in C\} \geq$

$\geq \inf\{d_1(x, z) + d_2(z, u) : z \in C\}$

and we got what we wanted.

19. Show that every dilation is one-to-one and uniformly continuous.

Solution. Let $f : (X, d_1) \to (Y, d_2)$ be a dilation of stretching factor α. It is obvious that f is one-to-one. Take any $\varepsilon > 0$, and choose $\delta = \dfrac{\varepsilon}{\alpha}$. Then if $d_1(x_1, y_1) < \delta$, we have $\alpha d_1(x_1, y_1) < \varepsilon$, which means that $d_2(f(x_1), f(y_1)) < \varepsilon$ and the claim is proven.

21. Let (X_1, d_1) and (X_2, d_2) be metric spaces. Show that if $f : X_1 \to X_2$ is a dilation, then $f^{-1} : f(X_1) \to X_1$ is continuous.

Solution. Suppose f is of stretching factor α. We use Proposition 3 to show that f^{-1} is continuous at every point. Let $\{f(x_i)\}$ be a sequence converging to $f(x)$ in $f(X_1)$. Now consider the sequence $\{x_i\}$ in relation to x. Take any $\varepsilon > 0$. There is an N, such that if $n > N$ we have $d_2(f(x_n), f(x)) < \alpha\varepsilon$. Since $d_2(f(x_n), f(x)) = \alpha d_1(x_n, x)$, it follows that for $n > N$, $\alpha d_1(x_n, x) < \alpha\varepsilon$, that is, $d_1(x_n, x) < \varepsilon$. So $\{x_i\}$ converges to x.

23. Let $f : \mathbb{R}^2 \to \mathbb{R}^2$ be a dilation. Prove the following:

    (a) f maps lines to lines.

    (b) f sends angles (pairs of rays emanating from a single point) to angles of equal size.

    (c) f maps circles to circles.

    (d) $\mathbb{R}^2$ is not a fractal.

[Note: Justifying that $\mathbb{R}^n$ is not a fractal for all $n \geq 3$ can be done in steps similar to the ones indicated in this exercise.]

Solution

    (a): Let f be of stretching factor α. Take two points $\mathbf{a}$ and $\mathbf{b}$ in $\mathbb{R}^2$, and let $\mathbf{x}$ be a point on the line passing through $\mathbf{a}$ and $\mathbf{b}$. Assuming $\mathbf{x}$ is between $\mathbf{a}$ and $\mathbf{b}$, we have $d(\mathbf{a}, \mathbf{x}) + d(\mathbf{x}, \mathbf{b}) = d(\mathbf{a}, \mathbf{b})$. Then $\alpha d(\mathbf{a}, \mathbf{x}) + \alpha d(\mathbf{x}, \mathbf{b}) = \alpha d(\mathbf{a}, \mathbf{b})$, and so $d(f(\mathbf{a}), f(\mathbf{x})) + d(f(\mathbf{x}), f(\mathbf{b})) = d(f(\mathbf{a}), f(\mathbf{b}))$, which tells us that $f(\mathbf{x})$ is on the line through $f(\mathbf{a})$ and $f(\mathbf{b})$. A similar argument could be made if $\mathbf{x}$ is not between $\mathbf{a}$ and $\mathbf{b}$.

    (b) This is true since similar triangles will be mapped to similar triangles.

    (c) That every point of a circle of radius r is mapped to a point on a circle of radius αr is obvious. That this is onto follows from (b).

    (d) We show that any dilation $f : \mathbb{R}^2 \to \mathbb{R}^2$ is onto, and hence $\mathbb{R}^2$ is not a fractal. Since a family of concentric circles of unbounded radii must map to a family of concentric circles of unbounded radii, the conclusion that f is onto follows.

Chapter 3: Topological Spaces: Definition and Examples

3.1 The Definition and Some Simple Examples.

Solutions of some exercises

---

1. Find all topologies over $X = \{1,2\}$.

Solution. $\tau_1 = \{\varnothing, X\}$, $\tau_2 = \{\varnothing, \{1\}, X\}$, $\tau_3 = \{\varnothing, \{2\}, X\}$, $\tau_4 = \{\varnothing, \{1\}, \{2\}, X\}$.

3. Show that the number of topologies over a finite set X grows exponentially with respect to number of elements of X. More precisely, show that there are at least 2^n topologies over a set with n elements.

Solution. We may assume $X = \{1, 2, ..., n\}$. For each sequence of length n containing only 0 or 1 construct a nested topology by interpreting the number in position i as "do or do not use the element number i in X" depending on whether that number is 1 or 0 respectively. For example, with $X = \{1,2,3,4\}$, the sequence $0,1,0,1$ gives rise to the nested topology $\{\varnothing, \{2,4\}, X\}$. That will give us $2^n - 1$ different nested topologies (the sequence $(0,0,...,0,0)$ and $(1,1,...,1,1)$ yield the same nested topology). We would then need one more topology to complete the justification.

5. Let X be a set, let Y be a space, and let $f : X \to Y$ be a mapping. Show that $\tau = \{f^{-1}(V) : V \underset{open}{\subseteq} Y\}$ is a topology over X.

Solution. (i) $\varnothing = f^{-1}(\varnothing) \in \tau$, $X = f^{-1}(Y) \in \tau$.

(ii) Let $U_i \in \tau$, $i \in I$. Then $U_i = f^{-1}(V_i)$, $i \in I$, where each V_i is open in Y. We have $\bigcup_{i \in I} U_i = \bigcup_{i \in I} f^{-1}(V_i) = f^{-1}\left(\bigcup_{i \in I} V_i\right)$, and the last set is in τ since $\bigcup_{i \in I} V_i$ is open in Y.

(iii) Let $U_i \in \tau$, $i = 1, 2, \ldots, n$. Then $U_i = f^{-1}(V_i)$, $i = 1, 2, \ldots, n$, where each V_i is open in Y. We have $\bigcap\limits_{i=1}^{n} U_i = \bigcap\limits_{i=1}^{n} f^{-1}(V_i) = f^{-1}\left(\bigcap\limits_{i=1}^{n} V_i\right)$, and the last set is in τ since $\bigcap\limits_{i=1}^{n} V_i$ is open in Y.

7. Show that the countable complement topology over a set X is indeed a topology.

Solution sketch: the argument used in Example 5 needs only minor modifications.

9. Let X be any (always non-empty) set, and let τ consist of a collection of subsets of X linearly ordered by $\subset$, together with the sets $\varnothing$ and X. Show that τ need not be a topology.

Solution. $X = \mathbb{Z}$, and set $U_1 = \{0\}$, $U_{2n} = \{0, 1, 2, \cdots, n\}$, $U_{2n+1} = \mathbb{Z}^{+} \cup \{-1, -2, \cdots, -n\}$, $n = 1, 2, \ldots$. Then the family of these sets together with $\varnothing$ and X is linearly ordered by $\subset$ but it is not a topology since the second axiom fails: $\bigcup\limits_{i=1}^{\infty} U_{2i}$ is not in τ.

11. Let τ_1 and τ_2 be two topologies over a set X. Show that $\tau_1 \cap \tau_2$ is also a topology over X. More generally, show that if $\{\tau_i : i \in I\}$ are topologies over X, then so is $\bigcap\limits_{i \in I} \tau_i$.

Solution. We verify the axioms (i), (ii) and (iii) in the definition of a topological space.

(i) is obvious.

(ii) Suppose U_j, $j \in J$, are such that $U_j \in \bigcap\limits_{i \in I} \tau_i$ for every $j \in J$. Then each U_j is in each τ_i. Hence $\bigcup\limits_{j \in J} U_j$ is also in each τ_i, and so it is in $\bigcap\limits_{i \in I} \tau_i$.

(iii) The argument in part (ii) needs only minor modifications.

12. Suppose X is a space, Y is a set, and $f : X \to Y$ is a mapping. Define a subset U of Y to be open if and only if $f^{-1}(U)$ is open in X. Show that this defines a topology over Y. Generalize: Let X_i be spaces, let Y be a set and let $f_i : X_i \to Y$, $i \in I$, be mappings. Show that $\mathcal{T} = \left\{ U \subset Y : \text{for every } i \in I, \, f_i^{-1}(U) \underset{open}{\subset} X_i \right\}$ is a topology over Y.

Solution. Again we verify the axioms (i), (ii) and (iii) in the definition of a topological space.

(i) is easy.

(ii) Suppose $\{U_j : j \in J\} \subset \mathcal{T}$. We need to show that $\bigcup_{j \in J} U_j \in \mathcal{T}$. The assumption means $f_i^{-1}(U_j)$ is open in X_i for every $i \in I$ and for every $j \in J$. Now,

$$f_i^{-1}\left(\bigcup_{j \in J} U_j\right) = \bigcup_{j \in J} f_i^{-1}(U_j)$$ for every i. Since the right-hand side is open in X_i, we showed that $\bigcup_{j \in J} U_j \in \mathcal{T}$.

(iii) is similar to (ii).

15. Let τ be the collection defined in Example 8, except that we do not require that X be in τ. Show that in that case τ is not necessarily a topology.

[Remark regarding the statement of the question: not requiring that X be in τ, does not mean that X is not in τ.]

Solution. Let $\mathbb{R}^* = \mathbb{R} \cup \{*\}$ where $*$ is not in $\mathbb{R}$. Extend the usual linear order over $\mathbb{R}$ by setting $*$ to be the largest element. Then $\mathbb{R}^*$ is not a union of open intervals generated by that linear ordering (as in Example 8).

3.2 Some Basic Notions

Solutions of some exercises
---

1. Consider the illustration 3.5 and its caption. Find a topology over $\mathbb{R}^2$ such that b is an interior point for S and at the same time a is not an interior point for S.

Solution. $\left\{\varnothing, \{b\}, \mathbb{R}^2\right\}$.

3. Show that finite unions of closed sets are closed. Give an example of a family of closed sets in $\mathbb{R}$ (usual topology), such that their union is not closed.

Solution. Let F_i, $i = 1, 2, \ldots, n$, be closed sets. Then $\left(\bigcup_{i=1}^{n} F_i\right)^c = \bigcap_{i=1}^{n} F_i^c$, which, being a finite intersection of open sets, is an open sets itself. Consequently $\bigcup_{i=1}^{n} F_i$ is closed.

For the second question, consider the family $\left\{\left[\dfrac{1}{j}, 1\right] : j = 1, 2, \ldots\right\}$ of closed intervals.

Their union is the semi-closed interval $(0, 1]$, which is not closed.

5. Find $\mathbb{Q}'$ and $(\mathbb{R} \setminus \mathbb{Q})'$ if $\mathbb{R}$ is equipped with
   (i)   discrete topology
   (ii) indiscrete topology
   (iii) finite complement topology.

Solution. (i) $\mathbb{Q}' = \varnothing = (\mathbb{R} \setminus \mathbb{Q})'$
         (ii) $\mathbb{Q}' = \mathbb{R} = (\mathbb{R} \setminus \mathbb{Q})'$
         (iii) $\mathbb{Q}' = \mathbb{R} = (\mathbb{R} \setminus \mathbb{Q})'$

7. Let A and B be subsets of a space X and let $A \subset B$ in parts (a), (b) and (c). Prove the following.
   (a) $A' \subset B'$.
   (b) $\text{int}(A) \subset \text{int}(B)$
   (c) $\bar{A} \subset \bar{B}$

(d) $(A \cup B)' = A' \cup B'$

Solution. (a) Assume $x \in A'$. This means that for every open neighborhood U_x of x, we have $U_x \cap (A \setminus \{x\}) \neq \varnothing$. Since $A \subset B$, this implies that $U_x \cap (A \setminus \{x\}) \subset U_x \cap (B \setminus \{x\})$, and so for every U_x, $U_x \cap (B \setminus \{x\}) \neq \varnothing$.

(b) Let $x \in \mathrm{int}(A)$. Hence there is a set U that is open in X and such that $x \in U \subset A$. Since $A \subset B$, we have $x \in U \subset A \subset B$, and hence $x \in \mathrm{int}(B)$.

(c) By Theorem 7, $\bar{A} = A \cup A'$. Since $A \subset B$, and by part (a), we have $A \cup A' \subset B \cup B'$. By Theorem 7, $B \cup B' = \bar{B}$. All this yields $\bar{A} \subset \bar{B}$.

(d) $\subset$: Suppose $x \notin A' \cup B'$. This last assumption implies that $x \notin A'$ and $x \notin B'$. Hence there exist an open neighborhoods U_x and V_x of x such that $U_x \cap (A \setminus \{x\}) = \varnothing$ and $V_x \cap (B \setminus \{x\}) = \varnothing$. Then $(U_x \cap V_x) \cap ((A \cup B) \setminus \{x\}) = \varnothing$, and so $x \notin (A \cup B)'$.

$\supset$: Suppose $x \in A' \cup B'$. Then $x \in A'$ or $x \in B'$. By symmetry we may assume $x \in A'$. By part (a) $A' \subset (A \cup B)'$, and so $x \in (A \cup B)'$.

9. Let X be a metric space and let A be a subset of X. Show that if $x \in A'$ then there is a sequence (x_n) of elements in A that are distinct from x such that $(x_n) \to x$. Show that that is not necessarily true in case X is not a metric space.

Solution. For every ball $B\left(x, \dfrac{1}{n}\right)$, $n = 1, 2, \ldots$, there is an element $x_n \neq x$, such that $x_n \in A$. Then $(x_n) \to x$.

For the second part, look at co-countable topology over $\mathbb{R}$: $0 \in \mathbb{R}'$, but there is no sequence converging to 0 (since the complement of the set containing the elements of a sequence is open).

11. Consider the usual space $\mathbb{R}$ and let $\mathbb{R}_1$ and $\mathbb{R}_2$ be two disjoint copies of that space. Define a topology over $\mathbb{R}_1 \cup \mathbb{R}_2$ by declaring open sets to be all of the sets of type $A_1 \cup A_2$ where A_1 and A_2 are copies in $\mathbb{R}_1$ and $\mathbb{R}_2$ respectively of a single open set A in $\mathbb{R}$.

(a) Show that this defines a topology over $\mathbb{R}_1 \cup \mathbb{R}_2$.

(b) Find a subset A of $\mathbb{R}_1 \cup \mathbb{R}_2$ such that $\partial(\partial A)$ is a proper subset of ∂A.

Solution for part (b). The subsets of $\mathbb{R}_1$ will come with subscript 1, those in $\mathbb{R}_2$ will come with subscript 2. Consider $A = \left((0,1)_1 \cap \mathbb{Q}_1\right) \cup \left((0,1)_2 \cap \mathbb{Q}_2\right)$. Then $\partial A = [0,1]_1 \cup [0,1]_2$ and $\partial\partial A = \{0,1\}_1 \cup \{0,1\}_2$.

14. Which of the following topologies over $\mathbb{Q}$ is Hausdorff?
      (a) Co-finite topology.
      (b) Co-countable topology.

Solution. Only (b).

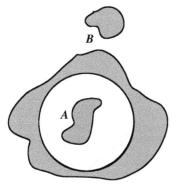

Illustration 3.12

15. Define a topology over $\mathbb{R}^2$ as follows. Open sets are the empty set, $\{(x, y) \in \mathbb{R} : x^2 + y^2 = 1\}$, and all sets of type $A \bigcup B$ where A is any subset of the open disk $\{(x,y) \in \mathbb{R}^2 : x^2 + y^2 < 1\}$ and B is any subset of the complement of that open disk that contains a ring $\{(x,y) \in \mathbb{R}^2 : 1 \le x^2 + y^2 \le r\}$ for some $r > 1$ (see Illustration 3.12). Call it the In-Out topology over $\mathbb{R}^2$.

      (a) Prove that the In-Out topology over $\mathbb{R}^2$ is indeed a topology.

      (b) Show that in the In-Out topology the boundary of the closed unit disk $D^2 = \{(x,y) \in \mathbb{R}^2 : x^2 + y^2 \le 1\}$ is the open unit disk $\{(x,y) \in \mathbb{R}^2 : x^2 + y^2 < 1\}$. What is the interior of D^2?

Solution of part (b): Every point in the open unit disk is a boundary point for the closed unit disk since a neighborhood of that point must contain points of the ring $\{(x,y) \in \mathbb{R}^2 : 1 \le x^2 + y^2 \le r\}$. On the other hand, no point on the unit circle is a boundary point since the unit circle is open as stipulated in the statement of the problem.
$$\text{int}(D^2) = \{(x,y) \in \mathbb{R}^2 : x^2 + y^2 = 1\}.$$

17. Show that for every two subsets A and B of a space X the following is true.
      (a) $\text{int}(A \cap B) = (\text{int}\, A) \cap (\text{int}\, B)$
      (b) $(\text{int}\, A) \cup (\text{int}\, B) \subset \text{int}(A \cup B)$ (and give an example where the inclusion is proper).
      (c) $\overline{A \cup B} = \bar{A} \cup \bar{B}$
      (d) $\overline{A \cap B} \subset \bar{A} \cap \bar{B}$

Solution of (a) *and* (c).
      (a) Take $x \in \text{int}(A \cap B)$. Then there is an open set U such that $x \in U \subset A \cap B$. Then $x \in U \subset A$ and $x \in U \subset B$, so that $x \in \text{int}(A) \cap \text{int}(B)$.

      Conversely, choose $x \in \text{int}(A) \cap \text{int}(B)$. So there are open sets U and V such that $x \in U \subset A$ and $x \in V \subset B$. It follows that $x \in U \cap V \subset A \cap B$, and since $U \cap V$ is open, we have that $x \in \text{int}(A \cap B)$.

      (c) Since $\bar{A} \cup \bar{B}$ contains $A \cup B$ and is closed, and since $\overline{A \cup B}$ is the smallest closed set containing $A \cup B$, it follows that $\overline{A \cup B} \subset \bar{A} \cup \bar{B}$.

Conversely, take $x \in \overline{A} \cup \overline{B}$. So $x \in \overline{A}$ or $x \in \overline{B}$. By symmetry we may suppose that $x \in \overline{A}$. By Theorem 7(b), this means that $x \in A$ or $x \in A'$. If $x \in A$, then $x \in A \cup B \subset \overline{A \cup B}$. If $x \in A'$, then (by Exercise 7(a)), $x \in (A \cup B)'$, and hence, by Theorem 7(b), $x \in \overline{A \cup B}$. Hence $x \in \overline{A} \cup \overline{B}$ implies that $x \in \overline{A \cup B}$, and we have $\overline{A} \cup \overline{B} \subset \overline{A \cup B}$.

19. Show that every sequence of infinitely many elements in a bounded subset of $\mathbb{R}$ contains a convergent subsequence.

Solution. Let (x_n) be a sequence with infinitely many elements. We may suppose that the elements of that sequence are in some interval $I_1 = [a,b]$. Choose $x_{n_1} \in I_1 \cap \{x_n : n \in 1, 2, ...\}$. Subdivide this interval into two equal subintervals, and choose I_2 to be any of the two subintervals that contains infinitely many points (at least one of them must be such). Choose $x_{n_2} \in I_2 \cap \{x_n : n \in 1, 2, ...\}$. Iterate ad infinitum. By Cantor's Nested Intervals Theorem (Theorem 2 in 2.2), $\bigcap_{n=1}^{\infty} I_n$ is a singleton $\{y\}$. It is now easy to see that $\left(x_{n_k}\right)_{k=1}^{\infty}$ converges to y.

22. (a) Find three distinct open subsets A, B, and C, of $\mathbb{R}^2$ such that $\partial A = \partial B = \partial C$.

(b) Find three pairwise disjoint subsets A, B, and C, of $\mathbb{R}^2$ such that $A \cup B \cup C = \mathbb{R}^2$ and such that $\partial A = \partial B = \partial C$.

(c) Find four disjoint subsets A, B, C and D, of $\mathbb{R}^2$ such that $A \cup B \cup C \cup D = \mathbb{R}^2$ and such that $\partial A = \partial B = \partial C = \partial D$.

(d) Find infinitely many disjoint subsets of $\mathbb{R}^2$ that share a common boundary.

Solution.

(a) Take A to be any non-empty proper open subset of $\mathbb{R}^2$ such that $\mathbb{R}^2 \setminus \overline{A} \neq \varnothing$ (for example, take A to be an open disk), take $B = \mathbb{R}^2 \setminus \overline{A}$, and take $C = A \cup B$.

(b) $A = \{(x,y) : y > 0\}$, $B = \{(x,y) : y < 0\}$, $C = \{(x,y) : y = 0\}$.

(c) $A = \{(x,y) : y > 0\}$, $B = \{(x,y) : y < 0\}$, $C = \{(x,y) : y = 0, x \text{ is rational}\}$, $D = \{(x,y) : y = 0, x \text{ is irrational}\}$.

(d) Denote $A_i = \mathbb{Q} \times \mathbb{Q} + (i,i)$, $i \in \mathbb{R} \setminus \mathbb{Q}$. Then for every i, $\partial A_i = \mathbb{R}^2$. Moreover, if $(\mathbb{Q} \times \mathbb{Q} + (i,i)) \cap (\mathbb{Q} \times \mathbb{Q} + (j,j)) \neq \varnothing$, for some $i, j \in \mathbb{R} \setminus \mathbb{Q}$, then $A_i = A_j$. Here is a proof: suppose $(q_1, q_2) + (i,i) = (q_3, q_4) + (j,j)$ and take any $(r,p) + (i,i) \in A_i$. Then

$(r,p)+(i,i)=(r,p)-(q_1,q_2)+(q_1,q_2)+(i,i)=\big((r,p)-(q_1,q_2)+(q_3,q_4)\big)+(j,j)$, and the last pair is clearly in A_j, so that $A_i \subset A_j$. The other inclusion follows by symmetry.

An alternative solution: take the upper open plane, the lower open plane, and the *x*-axis considered as $\mathbb{R}$, then look at the sets $\mathbb{Q}+\dfrac{\sqrt{2}}{n}$, $n \in \mathbb{Z}^+$. It is very easy to see that these are pairwise disjoint. The boundary of each of these mentioned sets is the whole *x*-axis.

Extras.

1. Give a criterion (and justify it) so that X equipped with the following topology is Hausdorff.

    (a) Co-finite topology
    (b) Co-countable topology
    (c) Order topology
    (d) Nested topology.

2. Denote by $\mathcal{T}$ the usual (Euclidean) topology over $\mathbb{R}$, and by J the set of irrational numbers. Define $\sigma = \{U \cup T \; ; \; U \in \mathcal{T}, T \subset J\}$.

    (a) Prove that σ is a topology over $\mathbb{R}$ and that $\mathcal{T} \subset \sigma$.

    (b) A point x in a space X is **isolated** if $\{x\}$ is open in X. Find all isolated points in $\mathbb{R}$ equipped with σ.

3.3 Bases

Solutions of the odd-numbered exercises
---

1. Consider the set $\mathcal{B} = \{\{(c,y) \in \mathbb{R}^2 : a < y < b\} : a,b,c \in \mathbb{R}, a < b\}$ of all vertical line segments in $\mathbb{R}^2$. Show that $\mathcal{B}$ is a basis for a topology over $\mathbb{R}^2$.

Solution. We use Theorem 2. (1) It is obvious that $\bigcup_{B \in \mathcal{B}} B = \mathbb{R}^2$. (2) Take

$B_1 = \{(c_1,y) : a_1 < y < b_1\} \in \mathcal{B}$ and $B_2 = \{(c_2,y) : a_2 < y < b_2\} \in \mathcal{B}$, and choose $x \in B_1 \cap B_2$. This forces $c_1 = c_2$. Then $x \in \{(c_1,y) : \max\{a_1,a_2\} < y < \min\{b_1,b_2\}\} \in \mathcal{B}$.

3. Show that the family $\mathcal{B}$ in the definition of the tangent disc topology is indeed a basis for a topology over $\mathbb{R}^2_{up}$.

Solution. Follow Theorem 2: (1) Obvious. (2) Take any $B_1, B_2 \in \mathcal{B}$ with non-empty intersection. If they share a common point on the x-axis, then $B_1 \cap B_2$ is either B_1 or B_2. Otherwise their intersection is the same as the intersection of the associated Euclidean open disk (B_i-s without the x-axis), and for every $x \in B_1 \cap B_2$ there is a Euclidean open disk B in the open upper half plane such that, $x \in B \subset B_1 \cap B_2$.

5. Show that the family $\mathcal{B}$ in the definition of the order topology (Example 9) is indeed a basis.

Solution. Follow Theorem 2: (1) follows from the axioms of a linear order. (2) Suppose $x \in (a,b) \cap (c,d)$. Notice that in all cases $(a,b) \cap (c,d)$ is also an interval of the type (e,f), where $e,f \in \{a,b,c,d\}$. So, $x \in (e,f) \subset (a,b) \cap (c,d)$. The cases involving semi-closed intervals are similar.

7. Show that the topology generated by a basis $\mathcal{B}$ is the weakest topology that contains the elements of $\mathcal{B}$.

Solution. It follows directly from the definition of topological spaces, that every topology containing $\mathcal{B}$ must also contain the unions of the elements of $\mathcal{B}$. Consequently, the topology generated by $\mathcal{B}$ must be contained in any other topology containing the sets of $\mathcal{B}$.

9. Show that if $\{\mathcal{B}_x : x \in X\}$ is a collection of local bases, then $\bigcup_{x \in X} \mathcal{B}_x$ is a basis for X.

Solution. We need to show that every open subset U is a union of the elements in $\bigcup_{x \in X} \mathcal{B}_x$.

For every $x \in X$ there is an element of $B_x \in \mathcal{B}_x$ such that $x \in B_x \subset U$. It follows that $U = \bigcup_{x \in U} U_x$, and we got what we wanted.

11. Let X be a second countable space. Show that every basis for X contains a countable subset that is also a basis for X.

Solution. Let $\mathcal{B}$ be a countable basis for the space X, and let C be any basis for the space X. Choose any $B \in \mathcal{B}$. Since C is a basis, B is a union of elements of C. Assume that the elements of C used in this union are $\{C_j : j \in J\}$. Since $\mathcal{B}$ is a basis, each of C_j is a union of elements of $\mathcal{B}$. Since $\mathcal{B}$ is countable, all of these members of $\mathcal{B}$, used to express as unions all C_j, $j \in J$, make a countable set $\{B_i : i = 1, 2, \dots\}$. We notice that for each $i = 1, 2, \dots$ there is $j \in J$ such that $B_i \subset C_j \subset B$. Now, for each $i = 1, 2, \dots$ choose exactly one $j \in J$ such that $B_i \subset C_j \subset B$. This gives us a countable set $\{C_j : j = 1, 2, \dots\}$ of elements in C. It follows from our construction that that $B = \bigcup_{j=1}^{\infty} C_j$. Repeat this procedure for every $B \in \mathcal{B}$. Since a countable union of countable sets is a countable set, it follows that the members of C that we have thus chose constitute a countable set $\mathcal{D}$. Since every member of $\mathcal{B}$ is a union of elements of $\mathcal{D}$, it easily follows that $\mathcal{D}$ is a countable basis for X. We achieved what we wanted.

13. Show that every metric space is first countable.

Solution. For every x in the metric space, the family $\left\{ B\left(x, \dfrac{1}{n}\right) : n = 1, 2, \dots \right\}$ is a countable local basis at x.

15. Let $\mathcal{P}$ be the set of all polynomials on n variables. For a polynomial $p(x_1, x_2, \dots, x_n) \in \mathcal{P}$, define $Z(p) = \{(x_1, x_2, \dots, x_n) \in \mathbb{R}^n : p(x_1, x_2, \dots, x_n) = 0\}$. Show that the collection $\mathcal{B} = \{\mathbb{R}^n \setminus Z(p) : p \in \mathcal{P}\}$ is a basis for a topology over $\mathbb{R}^n$. This is the **Zariski topology** over $\mathbb{R}^n$. Show that for $n = 1$ the Zariski topology coincides with the co-finite topology over $\mathbb{R}$.

Solution. Following Theorem 2. (1) is immediate, since every point in $\mathbb{R}^n$ lies outside the vanishing set of some polynomial. (2) Take $B_1 = \mathbb{R}^n \setminus Z(p_1)$ and $B_2 = \mathbb{R}^n \setminus Z(p_2)$ in $\mathcal{B}$. Then $B_1 \cap B_2 = \mathbb{R}^n \setminus Z(p_1 \cdot p_2)$, and it is also in $\mathcal{B}$.

In case $n = 1$, for every polynomial $p(x_1)$, the set $Z(p)$ is finite. Moreover, every finite subset of $\mathbb{R}$ is $Z(p)$ for some polynomial $p(x_1)$. Consequently the set $\mathcal{B}$ in this case consists of the complements of all finite sets, Hence the same is true for the set of all open sets, and thereby the topology is co-finite.

17. Let (X, τ_1) and (Y, τ_2) be topological spaces such that $X \cap Y = \varnothing$. Show that $\tau_1 \cup \tau_2$ is a basis for a topology over $X \cup Y$. Show that that this claim is in general false when X and Y are not disjoint.

Solution. Theorem 2: (1) is plain. (2) If $x \in B_1 \cap B_2$, where $B_1, B_2 \in \tau_1 \cup \tau_2$, then (since $X \cap Y = \varnothing$) either $B_1, B_2 \in \tau_1$ or $B_1, B_2 \in \tau_2$. In both cases $B_1 \cap B_2 \in \tau_1 \cup \tau_2$. For a counterexample in case when X and Y are not disjoint, take $X = \{1,2\}$, $Y = \{2,3\}$, and both τ_1 and τ_2 indiscrete. Notice that (2) fails for $\{2\} = X \cap Y$.

19. Let X be a space with a nested topology, and denote, as usual, the power set of X by $\mathcal{P}(X)$. For every two open subsets U, and V of X, denote

$$(U,V) = \left\{ W \underset{open}{\subseteq} X : U \subsetneq W \subsetneq V \right\}.$$

    (a) Show that the collection $\left\{ (U,V) : U, V \underset{open}{\subseteq} X \right\} \cup \{\varnothing\} \cup \{\mathcal{P}(X)\}$ is a basis for a topology over $\mathcal{P}(X)$. Call it the *power topology*.

    (b) Show that if $|X| \geq 2$, then the power topology over $\mathcal{P}(X)$ is not metrizable.

Solution.

    (a) We need to show that if $A \in (U_1, V_1) \cap (U_2, V_2)$, then there is (U_3, V_3) such that $A \in (U_3, V_3) \subset (U_1, V_1) \cap (U_2, V_2)$. The assumption that $A \in (U_1, V_1) \cap (U_2, V_2)$ implies that $U_1 \subsetneq A \subsetneq V_1$, and that $U_2 \subsetneq A \subsetneq V_2$. It is then easy to see that $(U_1 \cup U_2, V_1 \cap V_2)$ will do for (U_3, V_3).

    (b) Look at $\{\{x_1\}\}$ and $\{\{x_2\}\}$. Are there open sets U and V containing these two respectively and being disjoint? If so, there are such open sets from the above basis. It has to be that $U = (\varnothing, U_1)$ and $V = (\varnothing, V_1)$ since the empty set is the only proper subset of the singleton. But these two cannot be disjoint, since the sets U_1 and V_1 are open in the nested topology, so one is a subset of the other. Hence the power space is not Hausdorff, and so it is not metrizable

21. Suppose $f : X \to Y$ is a mapping from a set X into a space Y. Show that the set $\{f^{-1}(U) : U \text{ is open in } Y\}$ is a basis for some topology over X. Is the set $\{f^{-1}(U) : U \text{ is open in } Y\}$ a topology over X?

Solution.
Use Theorem 2. (1) It is true since X is in this set. (2) This follows from the observation that $f^{-1}(U \cap V) = f^{-1}(U) \cap f^{-1}(V)$.

The second question is answered in affirmative in Exercise 5, Section 3.1.

23.   (a) Show that if X is 1st countable and $x \in X$ is an accumulation point for a subset A of X, then there exists sequence $x_1, x_2, \ldots$ of elements in A that converges to x.
    (b) Show that the co-countable topology over $\mathbb{R}$ is not first countable.
    (c) Find a subset A of $\mathbb{R}$ equipped with co-countable topology, and a point $x \in \mathbb{R}$ such that $x \in A'$ but no sequence of elements of A converges to x.

Solution.
    (a) Suppose $\mathcal{B}_x = \{B_1, B_2, \ldots\}$ is a countable local basis at x. Then $B_1 \cap (A \setminus \{x\}) \neq \varnothing$. Choose $x_1 \in B_1 \cap (A \setminus \{x\})$. By the same argument $B_1 \cap B_2 \cap (A \setminus \{x\}) \neq \varnothing$, and so there is a point x_2 in that set. In general, we choose $x_n \in B_1 \cap B_2 \cap \ldots \cap B_n \cap (A \setminus \{x\})$, thus obtaining a sequence $x_1, x_2, \ldots$ of elements in A. This sequence converges to x, since if U is an open neighborhood of x, then there is a local basis element B_k such that $x \in B_k \subset U$, so that $x_k, x_{k+1}, \ldots$ are all in U.
    (b) Take $\mathbb{R}$ with the co-countable topology, and $A = (0,1)$. Then $1 \in A' (= \mathbb{R})$ but there is no sequence converging to 1, since $\mathbb{R} \setminus \{a_n\}$ is open and has an empty intersection with $\{a_n\}$. According to part (a), this tells us that $\mathbb{R}$ with the co-countable topology cannot be first countable.
    (c) Take A to be the set of irrationals and take x to be any rational number.

3.4 Dense, nowhere dense sets and related matters

Solutions of the odd-numbered exercises
---

1. Find a space X such that $X' = \varnothing$. True or false: if a space X is such that $X' \neq \varnothing$ then X' is dense in X?

Solution. For any discrete space X, $X' = \varnothing$.
      False! Take $X = \{1,2,3\}$ and $\mathcal{T} = \{\varnothing,\{1\},\{1.2\},X\}$. Then $X' = \{2,3\}$ and $\overline{\{2,3\}} = \{2,3\}$, and so X' is not dense in X.

3. Show that the Sorgenfrey line is Hausdorff but not metrizable.

Solution. It is obvious that the Sorgenfrey line is Hausdorff. If it was metrizable, then, by Proposition 3 and exercise 2, it would be second countable. That would contradicts Example 15 in Section 3.3.

5. Find reasonable sufficient and necessary conditions for a space equipped with the countable complement topology to be separable.

Solution. We prove that such a space X is separable iff it is countable.
      $\Leftarrow$: Suppose X is countable. Then X itself is a countable dense set. Hence X must be separable.
      $\Rightarrow$: Suppose X is uncountable. Take any countable set D. Then the complement D^c of D in X is not empty and it is open in the co-countable topology. Since it is clearly disjoint from S, it follows that it could not be dense in X. Hence X is not second countable.

7. Show that the only closed and dense subset of a space X is X itself.

Solution. $D = \overline{D} = X$.

9. Show that a finite union of nowhere dense sets is nowhere dense.

Solution. Let A_i, $i = 1, 2, \ldots, n$, be nowhere dense sets. Hence $\mathrm{int}(\overline{A_i}) = \varnothing$ for every i.

Assume $\mathrm{int}\left(\overline{\bigcup_{i=1}^{n} A_i}\right) \neq \varnothing$. By exercise 17 (c), Section 3.2, $\overline{\left(\bigcup_{i=1}^{n} A_i\right)} = \bigcup_{i=1}^{n} \overline{A_i}$. Since we have

assumed its interior is not empty, there is a non-empty open set U_1 such that $U_1 \subset \bigcup_{i=1}^{n} \overline{A_i}$.

If $U_1 \subset \overline{A_1}$, then we have a contradiction. So $U_2 = \overline{A_1}^c \cap U_1 \neq \varnothing$ and $U_2 \subset \bigcup_{i=2}^{n} \overline{A_i}$. Repeat

this argument finitely many times, eventually ending with $U_n \subset \overline{A_n}$, which gives us a
contradiction.

11.   (a) Show that A is nowhere dense in X if and only if for every non-empty open
subset U of X, there is an non-empty open set V such that $V \subset U$ and $V \cap \overline{A} = \varnothing$.
    (b) Show that A is nowhere dense in a metric space X if and only if for every non-
empty open subset U of X, there is a ball B such that $B \subset U$ and such that $\overline{B} \cap A = \varnothing$.

Solution.
    (a) Suppose A is nowhere dense in X. So the closure of A has no interior points.
Now take any open set U in X. If $U \cap \overline{A} = \varnothing$ we can take $V = U$ and we are done.
Otherwise, there is $x \in U \cap \overline{A}$. This x is certainly not an interior point for $\overline{A}$, and so there
is a point y in U out of $\overline{A}$. But then $y \in U \cap \left(\overline{A}\right)^c = V$, and we have our non-empty open
subset of U avoiding $\overline{A}$. Conversely, suppose $\mathrm{int}(\overline{A}) \neq \varnothing$; choose $x \in \mathrm{int}(\overline{A})$. Then there
is an open neighborhood U of x that is entirely in $\overline{A}$. Then obviously, each non-empty
part of U will also be in $\overline{A}$ and will have a non-empty intersection.
    (b) Since every open set contains an open ball within it, this part is a consequence
of part (a).

13. Show that if A is nowhere dense in X, then the complement $\left(\overline{A}\right)^c$ is dense in X.

Solution. Since $\overline{A}$ contains no non-empty open sets, the complement must intersect each
of them.

3.5 Continuous Mappings

Solutions of the odd-numbered exercises

---

1. (a) Suppose X is equipped with the discrete topology and let Y be any topological space. Show that every mapping $f : X \to Y$ is continuous.

  (b) Suppose Y is equipped with the indiscrete topology. Show that every mapping $f : X \to Y$ (X is any topological space) is continuous.

Solution.

  (a) Every subset of X is open. Hence, for every open $V \subset Y$, $f^{-1}(V)$ is open in X.

  (b) $f^{-1}(Y) = X$ and $f^{-1}(\varnothing) = \varnothing$ are both open in X.

3. Suppose $f : X \to Y$ is continuous and onto.

  (a) Show that if D is a dense subset of X then $f(D)$ is a dense subset of Y.

  (b) Show that if X is separable then so is Y.

Solution.

  (a) Let U be any non-empty open subset of Y. We show that $f(D) \cap U \neq \varnothing$. Since f is onto, $f^{-1}(U)$ is not empty, and since f is continuous, $f^{-1}(U)$ is open in X. Hence $f^{-1}(U) \cap D \neq \varnothing$. So $f\big(f^{-1}(U) \cap D\big) \neq \varnothing$. On the other hand $f\big(f^{-1}(U) \cap D\big) \subset f\big(f^{-1}(U)\big) \cap f(D) = U \cap f(D)$. Hence $U \cap f(D) \neq \varnothing$.

  (b) Let X be separable. So, there is a countable set D that is dense in X. Then $f(D)$ is also countable. By part (a) the set $f(D)$ is dense in Y.

5. Show that if $f : X \to Y$ is a bijection than f is open if and only if it is closed.

Solution. $\Rightarrow$ Suppose f is open and let F be an closed subset in X. Since f is a bijection, we have that $f(F) = \big(f(F^c)\big)^c$. Since f is an open mapping, $f(F^c)$ is an open set. Consequently $\big(f(F^c)\big)^c$ is a closed set, hence $f(F)$ is also closed. $\Leftarrow$ To justify this it suffices to use almost an identical argument.

7. (a) Let X be a space with topology τ, let p be a point not in X, and denote $Y = X \cup \{p\}$. Show that $\tau \cup \{Y\}$ is a topology over Y.

  (b) Find a mapping $f : X \to Y$ that is open but not closed.

Solution.

(a) It is obvious that all three axioms are satisfied.

(b) Consider the inclusion $in : X \to Y$ Obviously open sets go to open sets. On the other hand X is closed in X, but $X = f(X)$ is not closed in Y (since $\{p\}$ is not open in Y).

9. Let $f : X \to Y$ be a homeomorphism and let A be a subset of X. Show the following.

(i) If X is first countable, then so is Y.

(ii) $f(\operatorname{int} A) = \operatorname{int} f(A)$

(iii) $f(\bar{A}) = \overline{f(A)}$

(iv) $f(A') = (f(A))'$

Solution.

(i) If $y \in Y$ and $\mathcal{B}_x$ is a countable local basis at $x = f^{-1}(y)$, then $\{f(B) : B \in \mathcal{B}_x\}$ is a countable local basis at y.

(ii) Since $\operatorname{int}(A) \subset A$, it follows that $f(\operatorname{int}(A)) \subset f(A)$. Since $f(\operatorname{int}(A))$ is open, it follows from Proposition 2, Section 3.2, that $f(\operatorname{int}(A)) \subset \operatorname{int}(f(A))$. To prove the converse inclusion, start with $y \in \operatorname{int}(f(A))$. Then there is an open set V such that $y \in V \subset f(A)$. Denote $x = f^{-1}(y)$ and $U = f^{-1}(V)$. Since f is a homeomorphism, U is open, and $x \in U \subset A$. Consequently $x \in \operatorname{int}(A)$, and so $y = f(x) \in f(\operatorname{int}(A))$. Hence $f(\operatorname{int} A) \subset \operatorname{int} f(A)$.

(iii) By Theorem 7, Section 3.2, $\bar{A} = A \cup A'$. Hence $f(\bar{A}) = f(A \cup A')$, and, since f is a bijection, $f(A \cup A') = f(A) \cup f(A')$. By part (iv), $f(A') = (f(A))'$, and by Theorem 7, 3.2, $f(A) \cup (f(A))' = \overline{f(A)}$.

(iv) Take any $y \in f(A')$. Hence $y = f(x)$ for some $x \in A'$. Take any open set V around y. Then $U = f^{-1}(V)$ is an open neighborhood around x. Since $x \in A'$, there must be $z \in U \cap A$, $z \neq x$. Then $f(z) \in f(U \cap A) = f(U) \cap f(A) = V \cap f(A)$, and $f(z) \neq f(x) = y$. We showed that for every open neighborhood V of y,

$V \cap f(A) \setminus \{y\} \neq \varnothing$, thus showing that $y \in (f(A))'$. Hence $f(A') \subset (f(A))'$.

Conversely, suppose $y \in (f(A))'$. Then for every open neighborhood V of y, there is $z \in V \cap f(A)$, $z \neq y$. Then $f^{-1}(z) \in f^{-1}(V \cap f(A)) = f^{-1}(V) \cap A$, and $f^{-1}(z) \neq f^{-1}(y)$. Since f is a homeomorphism, the sets of type $f^{-1}(V)$ make the class of all open neighborhoods of $f^{-1}(y)$. Consequently $f^{-1}(y) \in A'$. Hence $y \in f(A')$. We showed that $f(A') \supset (f(A))'$, and hence completed the proof.

11. Let (X, d) be a metric space and let $x \in X$. Show that the mapping $f : X \to \mathbb{R}$ defined by $f(y) = d(x, y)$ is continuous.

Solution. We showed in Example 3, Section 2.2, that $d : X^2 \to \mathbb{R}$ is continuous. Since restrictions (to metric subspaces) and compositions of continuous mappings are continuous, and since $f = \left(d \big|_{\{x\} \times X} \right) \circ in$, where $in : X \to \{x\} \times X$ is the obvious mapping, it follows that f is also continuous.

13. Let A be the subset of $\mathbb{R}^2$ consisting of the black points in Figure 6.

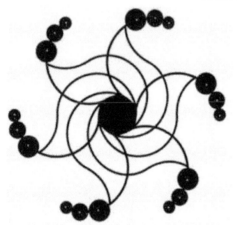

(a) Let A be equipped with the discrete topology. Is that space homeomorphic to the discrete space over $\mathbb{R}^2$?

(b) Let Y be the topological space over A with open sets being the empty set and the intersections of A with open disks centered at the center of A. Let Z be the topological space over $\mathbb{R}^2$ consisting of the empty set and all open disks centered at a fixed point. (Why are Y and Z indeed topological spaces?) Prove that the spaces Y and Z are not homeomorphic.

Illustration 3.26.

Solution. (a) Any two discrete spaces of equal cardinality are homeomorphic.

(b) The largest circle C that has a non-empty intersection with A intersects A at finitely many points. This finite set B is the boundary of the intersection with A of the open disk bounded by C. On the other hand the boundary of any non-empty open set in Z contains infinitely many points (on a circle, or all of $\mathbb{R}^2$). Existence of a homeomorphism between Y and Z would then contradict part (v) of Proposition 5.

15. Show that if $f : X \to \mathbb{R}$ and $g : X \to \mathbb{R}$ are continuous functions, then the function $h : X \to \mathbb{R}$ defined below is also continuous:

(a) $h(x) = |f(x)|$

(b) $h(x) = \min\{f(x), g(x)\}$

(c) $h(x) = \max\{f(x), g(x)\}$

Solution.

(a) Since $abs : \mathbb{R} \to \mathbb{R}$ defined by $x \mapsto |x|$ is continuous, and since h is the composition $abs \circ f$, it follows that h is also continuous.

(b) $\min\{f(x), g(x)\} = \dfrac{|f(x) + g(x)|}{2} - \dfrac{|f(x) - g(x)|}{2}$.

(c): $\max\{f(x), g(x)\} = \dfrac{|f(x) + g(x)|}{2} + \dfrac{|f(x) - g(x)|}{2}.$

17. Show that the fixed-point property (every continuous $f : X \to X$ has a fixed point) is a topological property.

Solution: Let X have a fixed point property, let $h : X \to Y$ be a homeomorphism, and let $g : Y \to Y$ be continuous. Then $h^{-1} \circ g \circ h : X \to X$ is also continuous. Hence there is $x \in X$ such that $h^{-1} \circ g \circ h(x) = x$. It follows that $g(h(x)) = h(x)$, and so g has a fixed point property too.

19.   (a) Show that if $f : X \to X$ is continuous and satisfies $f \circ f = identity$ then f is a homeomorphism.

(b) Show that if $f : \mathbb{R}^n \to \mathbb{R}^n$ is a homeomorphism without fixed points, then for every $x \in \mathbb{R}^n$ there is an open neighbourhood U of x such that $f(U) \cap U = \varnothing$.

Solution.

(a) If $g \circ h$ is a bijection, then h is 1-1, and g is onto. The result now follows, sice the identity mapping is a bijection.

(b) Suppose otherwise: this means that there exists $\mathbf{a} = (a_1, a_2, ..., a_n) \in \mathbb{R}^n$ such that for every open neighbourhood U of $\mathbf{a}$, we have $f(U) \cap U \neq \varnothing$. Consider the sequence of open n-cubes $S_m = \{(x_1, x_2, ..., x_n) : a_i - \dfrac{1}{m} < x_i < a_i + \dfrac{1}{m}, i = 1, 2, ..., n\}$,

$m = 1, 2, ...$ Notice that $\bigcap\limits_{m=1}^{\infty} S_m = \{\mathbf{a}\}$; so $\bigcap\limits_{m=1}^{\infty} f(S_m) = f\left(\bigcap\limits_{m=1}^{\infty} S_m\right) = \{f(\mathbf{a})\}$, where the first

equality is true since f is a homeomorphism, while the second is obvious. If $\mathbf{a}$ is in each $f(S_m)$, then it is in the intersection. But that intersection is $\{f(\mathbf{a})\}$, and so in that case $f(\mathbf{a}) = \mathbf{a}$ contradicting our assumption that f has no fixed points. So, there is S_m such that $\mathbf{a} \notin f(S_m)$.

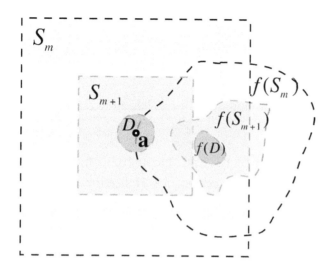

Figure 2. The point **a** could be on the boundary of $f(S_m)$, but it stays away from $\overline{f(S_{m+1})}$.

Now consider S_{m+1}: since f is a homeomorphism, $f(\overline{S}_{m+1}) \subset f(S_m)$. So **a** is out of $f(\overline{S}_{m+1})$ and so its distance to $f(\overline{S}_{m+1}) \, (= \overline{f(S_{m+1})})$ is positive. This implies that there is an open disk D around **a** and in S_{m+1}, such that D is disjoint from $f(S_{m+1})$. Hence D and $f(D) \subset f(S_{m+1})$ are disjoint, contradicting our assumption in the first sentence in this proof. With that the exercise is solved.

21. Consider the space $\mathbb{R}$ with the co-countable topology, and the space $\mathbb{Z}$ with co-finite topology. Then $\mathbb{R}$ is not first countable (Exercise 22 (b) in 3.3).

  (a) Show that the only convergent sequences in $\mathbb{R}$ are of type
  $x_1, x_2, \ldots, x_n, x, x, x, \ldots$ (i.e., sequences that eventually stabilize at a constant).

  (b) Define $f : \mathbb{R} \to \mathbb{Z}$ by $f(x) = \lfloor x \rfloor$ (the largest integer $\leq x$). Show that if (x_j) is a convergent sequence in $\mathbb{R}$, then $(f(x_j))$ is a convergent sequence in $\mathbb{Z}$.

  (c) Show that f is not continuous.

(With this exercise we establish that the first countability condition in Proposition 2 is necessary.)

Solution.

  (a) Let (x_j) converge to some $x \in \mathbb{R}$. Denote by A the set of all members in the sequence except x. Then A is countable, so $\mathbb{R} \setminus A$ is open neighbourhood of x. So, from some point on all members of the sequence must be there. But the only possible member of the sequence that is there is x.

  (b) This is obvious, since eventually constant sequences are mapped to eventually constant sequences, and these converge in $\mathbb{Z}$.

  (c) $U = \mathbb{Z} \setminus \{1\}$ is open in $\mathbb{Z}$. But $f^{-1}(U) = \mathbb{R} \setminus [1, 2)$ is not open in $\mathbb{R}$.

23. Let X be a space, let Y be a set and let $f_i : X \to Y$, $i \in I$, be mappings. Show that

$$\mathcal{T} = \left\{ U \subset Y : \text{for every } i \in I,\ f_i^{-1}(U) \underset{open}{\subset} X \right\} \text{ is the topology over } Y \text{ induced by}$$

$\{ f_i : i \in I \}$.

Solution. That $\mathcal{T}$ is a topology was noted in Exercise 15, in 3.1. We need to show that it is the largest topology making all f_i open. That it makes all f_i continuous is obvious. It is also obvious that no other subset of Y could be open, for else it would make some f_i discontinuous.

Chapter 4: Subspaces, Quotient Spaces, Product Spaces and More

4.1 Subspaces

Solutions of the odd-numbered exercises
---

1. Let Z be a space and let $X \subset Y \subset Z$. Show that the subspace X of Z is the same as the subspace X of the subspace Y of Z. (This exercise shows that it does not matter if we consider X as a subspace of Y or as a subspace of Z.)

Solution. Let U be an open set in the subspace X of Z. Then $U = V \cap X$ for some open subset V of Z. Now, $W = V \cap Y$ is open in Y. Then $W \cap X = (V \cap Y) \cap X = V \cap (Y \cap X) = V \cap X$. Hence $U = W \cap X$, and thus it is an open set in the subspace X of Y.

Conversely, suppose U is open in the subspace X of Y. Then $U = V \cap X$ for some open subset V of Y. Since Y is a subspace of Z, there is some W that is open in Z and such that $V = W \cap Y$. Then $U = (W \cap Y) \cap X = W \cap (Y \cap X) = W \cap X$, and hence it is open in X considered as a subspace of Z.

3. Show that the $(0,1) \cong \left(-\dfrac{\pi}{2}, \dfrac{\pi}{2} \right) \cong \mathbb{R}$ (usual topology). This shows that a proper subspace of a space X can be homeomorphic to X.

Solution. $f(t) = -\dfrac{\pi}{2}(1-t) + \dfrac{\pi}{2}t$ establishes a homeomorphism $(0,1) \rightarrow \left(-\dfrac{\pi}{2}, \dfrac{\pi}{2} \right)$.

$g(t) = \tan t$ is a homeomorphism $\left(-\dfrac{\pi}{2}, \dfrac{\pi}{2} \right) \rightarrow \mathbb{R}$.

5. Show that if Z is a subspace of Y and if $f : X \rightarrow Z$ is continuous, then $g : X \rightarrow Y$ defined by $g(x) = f(x)$, for every $x \in X$, is also continuous. Show that this statement is not always true if the word "continuous" is replaced by "open" (or "closed").

Solution. Let $f : X \rightarrow Z$ be continuous and let U be an open set in Y. Since $g(X) \subset Z$, it follows that $g^{-1}(U) = f^{-1}(U \cap Z)$. This last set is open since f is continuous.

The identity $id : [0,1) \to [0,1)$ is both open and closed (it is a homeomorphism), but the inclusion $in : [0,1) \to \mathbb{R}$ is neither open nor closed.

7. If X is a subspace of Y, show that the inclusion mapping $i : X \to Y$ (defined by $i(x) = x$, for every $x \in X$) is continuous.

Solution. Let U be open in Y. Then, since $i^{-1}(U) = U \cap X$ is open in X, i is continuous.

9. Show that if $\mathcal{B}$ is a basis for X and A is a subspace of X, then the set $\{B \cap A : B \in \mathcal{B}\}$ is a basis for the subspace A.

Solution. Let U be an open set in A. then $U = V \cap A$ for some open subset V of X. Since $\mathcal{B}$ is a basis for X, $V = \bigcup\limits_{\substack{i \in I \\ B_i \in \mathcal{B}}} B_i$. Then $U = V \cap A = \left(\bigcup\limits_{\substack{i \in I \\ B_i \in \mathcal{B}}} B_i \right) \cap A = \bigcup\limits_{\substack{i \in I \\ B_i \in \mathcal{B}}} (B_i \cap A)$, and we have what we claimed.

11. Let X be the subset of $\mathbb{R}^2$ defined by the (uncountably many) black points in Illustration 4.3.

(a) Show that X (as a subspace of $\mathbb{R}^2$) is not homeomorphic to $\mathbb{R}^2$.

(b) Consider $Y = \mathbb{R}^2$ with the topology induced by the post office metric (Section 2.1), with the designated point (the post office) being a white point in $\mathbb{R}^2$. Show that in that case X (as a subspace of Y) is homeomorphic to $\mathbb{R}^2$ equipped with the discrete topology.

Illustration 4.3

Solution.

(a) Notice that the smallest (middle) disk in X is open, and so is its complement in X. Hence X is a union of two non-empty open sets. If X and $\mathbb{R}^2$ were homeomorphic, then $\mathbb{R}^2$ would also be a union of two open sets. We now show that this is not possible (we will cover this theme more thoroughly in Chapter 6). Suppose $\mathbb{R}^2 = U \cup V$, where U and V are non-empty and open. Choose $u \in U$ and $v \in V$, and consider the line segment $L = \{u(1-t) + vt : t \in [0,1]\}$. By the least upper bound property of $\mathbb{R}$ there exists $t_0 = \sup\{t \in [0,1] : u(1-t) + vt \in U\}$. It is now easy to see that the openness of U and V prevents $u(1-t_0) + vt_0$ to be either in A or in B. This contradicts $\mathbb{R}^2 = U \cup V$.

(b) This follows from Exercise 5(a) in Section 2.1.

13. Show that every subspace of a Hausdorff space is Hausdorff.

Solution. Let A be a subspace of a Hausdorff space X, and let $u,v \in A$, $u \neq v$. Since X is Hausdorff there are open subsets U and V of X such that $u \in U$, $v \in V$, $U \cap V = \varnothing$. Then $u \in U \cap A$, $v \in V \cap A$, $(U \cap A) \cap (V \cap A) = \varnothing$, proving that A is Hausdorff.

15.   (a) Equip $X = \{a_1,a_2,a_3\}$ with the topology generated by the trivial metric d_t (Section 2.1). Find and visualize an embedding f from the space X into $\mathbb{R}^2$ such that $d_t(x,y) = d(f(x),f(y))$ for every $x,y \in X$ (d stands for the Euclidean metric).
         (b) Suppose $X = \{a_1,a_2,a_3,...,a_n\}$ with the topology generated by the trivial metric (Section 2.3). Find an embedding f from the space X into $\mathbb{R}^{n-1}$ such that $d_t(x,y) = d(f(x),f(y))$ for every $x,y \in X$.

Solution (b) and (a). Define $X \to \mathbb{R}^n$ by $a_i \mapsto (0,0,...,0, \frac{1}{\sqrt{2}},0,...,0)$, with the non-zero number at the i-th coordinate. (In case $n = 4$ this is a tetrahedron.) We have n points in $\mathbb{R}^n$ (each two at distance 1): they lie on an $(n-1)$-dimensional hyperplane. Now rotate this hyperplane so that it coincides with $\mathbb{R}^{n-1}$ (considered as the subspace of $\mathbb{R}^n$ obtained by deleting the last coordinate).

17. Let $X = A \cup B$ and suppose $U \subset A \cap B$ is open (closed) in both A and B. Show that U is open (closed, respectively) in X.

Solution. Suppose U is open in both A and B. So $U = U_1 \cap A = U_2 \cap B$ where both U_1 and U_2 are open in X. Then $U = U_1 \cap U = U_1 \cap (U_2 \cap B) = (U_1 \cap U_2) \cap B$, and, by symmetry $U = (U_1 \cap U_2) \cap A$. So,
$U = [(U_1 \cap U_2) \cap A] \cup [(U_1 \cap U_2) \cap B] = (U_1 \cap U_2) \cap (A \cup B) = (U_1 \cap U_2) \cap X$, so that U is open in X. Closeness could be done similarly.

4.2 Quotient Spaces

Solutions of the odd-numbered exercises

---

1. Give a sufficient and necessary criterion for the quotient maps to be homeomorphisms.

Solution. $q : X \to X/_{\sim}$ is a homeomorphism if and only if $\sim$ is the trivial equivalence ($x \sim y$ if and only if $x = y$).

3. Find an equivalence relation $\sim$ on $\mathbb{R}^2$ such that
    (a) $\mathbb{R}^2/_\sim$ is (homeomorphic to) a circle (equipped with the usual topology).
    (b) $\mathbb{R}^2/_\sim$ is (homeomorphic to) a sphere (equipped with the usual topology).

Solution.
    (a) The equivalence $\sim$: first project $\mathbb{R}^2$ into a vertical line and then wrap around the circle. Alternatively, consider $f(x, y) = (\cos 2\pi y, \sin 2\pi y)$ as a continuous mapping from $\mathbb{R}^2$ onto the unit circle, and the associated equivalence relations $\sim_f$ as defined in Proposition 2. Then $\mathbb{R}^2/_{\sim_f}$ is homeomorphic to the unit circle.)
    (b) First project each point on a ray emanating from the origin and outside the unit disk onto the point in the intersection of that ray with the unit circle, then identify a boundary point in the upper half-circle with the point straight below it on the lower half-circle. (Alternatively, by Proposition 2 it suffices to find an open and or closed subjective mapping $f : \mathbb{R}^2 \to S^1$: consider $f(x, y) = (\cos x \sin y, \sin x \sin y, \cos y)$ as a continuous mapping from $\mathbb{R}^2$ onto the unit sphere. Then $\mathbb{R}^2/_{\sim_f}$ is homeomorphic to the unit sphere.)

5. Show that the quotient mapping $q : X \to X/_{\sim}$ is closed if and only if for every closed $A \subset X$, the set $\bigcup_{[x] \cap A \neq \varnothing} [x]$ is closed in X. Show that the same is true if 'closed' is replaced by 'open'. [Hint: Proposition 1]

Solution of the first part.
Let A be a closed subset of X. Denote $F = \{[x] : [x] \cap A \neq \varnothing\}$. Notice that
$$\bigcup_{[x] \cap A \neq \varnothing} [x] = \bigcup_{[x] \in F} [x] \quad \text{and that } q(A) = F. \text{ By Proposition 1 (A subset } F \text{ of } X/_{\sim} \text{ is closed in}$$

$X/_\sim$ if and only if the set $\bigcup_{[x]\in F} [x]$ is closed in X), $\bigcup_{[x]\cap A\neq\varnothing} [x]$ is closed in X if and only if

F is closed in $X/_\sim$. Since q is closed if and only if $F = q(A)$ is closed for every closed subset A of X, the claim of the exercises follows.

7. Suppose $\sim$ is an equivalence relation over a space X, and suppose $f : X \to Y$ is a continuous mapping such that $f(x_1) = f(x_2)$ for every $x_1, x_2 \in X$ such that $x_1 \sim x_2$. Show that the mapping $g : X/_\sim \to Y$ defined by $g([x]) = f(x)$ is well defined and continuous.

Solution. That it is well defined is obvious. Let U be open in Y. Then $f^{-1}(U)$ is open in X. Now, it is easy to check that $f^{-1}(U) = p^{-1}(g^{-1}(U))$, where $p : X \to X/_\sim$ is the quotient mapping. It follows from the observation following Proposition 3 (or directly from the definition of the quotient topology) that $g^{-1}(U)$ is open in $X/_\sim$, and so g is continuous.

9. Consider the upper half-plane $\mathbb{R}^2_{up} = \{(x,y) : y \geq 0\}$ equipped with the tangent-disk topology. Partition $\mathbb{R}^2_{up}$ into equivalence classes, one of which consists of all elements of the x-axis, while the others are singletons; denote the resulting equivalence relation by $\sim$.

    (i) Is the sequence $\left\{ ((-1)^n, \frac{1}{n}) : n = 1,2,... \right\}$ convergent in $\mathbb{R}^2_{up}$? Justify your answer.

    (ii) Find a countable local basis at $[(0,0)]$, or show such a local basis does not exist. [Remark: $[(0,0)]$ is the equivalence class of $(0,0)$.]

Solution.

    (i) No. Any open (in $\mathbb{R}^2_{up}$) ball tangential to $(0,0)$ (containing that point) and of radius < 1 does not contain elements from the sequence.

    (ii) We show that a countable local basis at $[(0,0)]$ does not exist. Let $\{B_n : n = 1,2,...\}$ be any countable family of open (in $\mathbb{R}^2_{up}/_\sim$) neighborhoods of $[(0,0)]$. Let $q : \mathbb{R}^2_{up} \to \mathbb{R}^2_{up}/_\sim$ be the quotient map. Then each $q^{-1}(B_n)$ is an open neighborhood of the x-axis. Consequently, for each point $(x,0)$ on the x-axis there is an open (in $\mathbb{R}^2_{up}$) semi-disk D_x centered at that point and contained in $q^{-1}(B_n)$. Look at the points $(m,0)$, $m = 1,2,...$ on the x-axis. For each such point $(m,0)$ choose a semi-disk D_m within $q^{-1}(B_m)$. Let r_m be the radius of the disk D_m, and let D_m^* be the open half disk of radius

$^r m/_2$. Consider the set $U = \left(\bigcup_{m=1}^{\infty} D_m^* \right) \cup \left(\bigcup_{x \neq m} D_x \right)$: it is an open neighborhood of the x-axis.

The set $q(U)$ is then an open neighborhood of $[(0,0)]$ in $^{\mathbb{R}^2_{up}}/_{\sim}$. It is then readily visible that no B_n is a subset of $q(U)$, and so $\{B_n : n = 1,2,...\}$ is not a local basis.

11. Define an equivalence relation $\sim$ on $\mathbb{R}$ as follows: $x \sim y$ if and only if either $x = y$ or $x,y \in [0,1)$. Show that $^{\mathbb{R}}/_{\sim}$ is not homeomorphic to $\mathbb{R}$. Show that the same conclusion holds if $[0,1)$ is replaced by $(0,1]$ or by $(0,1)$.

Solution. The interval $[0,1)$ is one element in $^{\mathbb{R}}/_{\sim}$. Every set around $[0,1)$ that is open in $^{\mathbb{R}}/_{\sim}$ must contain the point $1 \in {}^{\mathbb{R}}/_{\sim}$. So $^{\mathbb{R}}/_{\sim}$ fails to be Hausdorff. As Hausdorff-ness is a topological property, $^{\mathbb{R}}/_{\sim}$ is not homeomorphic to $\mathbb{R}$. Similar argument can be used to justify the last part of this exercise.

13. Let $\sim$ be the smallest relation on $\mathbb{R}$ defined by $x \sim y$ for every integers x and y. Then the quotient space $^{\mathbb{R}}/_{\sim}$ is a wedge of countably many circles (countably many circles with one common point).

    (a) Show that $^{\mathbb{R}}/_{\sim}$ is not embeddable in $\mathbb{R}^2$.

    (b) Show that $^{\mathbb{R}}/_{\sim}$ is not first countable (hence, neither is it second countable

Solution. (a) This follows from (b) and the fact that first countability is both hereditary (Proposition 4 in 4.1) and topological (preserved under homeomorphisms).

    (b) Denote the equivalence class of integers by z. So $z \in {}^{\mathbb{R}}/_{\sim}$. We show that there is no countable local basis at z. Suppose otherwise, and denote this countable local base by $\mathcal{B}_z = \{B_s : s = 1,2,...\}$. Since the basis sets are open, not of them is $\{z\}$ (which is not open). Hence for every s, there is $x_s \in B_s \setminus \{z\}$. By Exercise 6 in 3.3 the sequence (x_s) must converge to z. On the other hand $U = \left({}^{\mathbb{R}}/_{\sim} \right) \setminus \{x_s : s = 1,2,...\}$ is an open neighborhood of z that contains no elements from (x_s) . This is a contradiction.

15. Define $\sim$ on $\mathbb{R}$ as follows: $x \sim y$ if $|x - y| \in \mathbb{Q}$. Confirm that this is an equivalence relation and describe the open sets in the quotient space.

Solution. The first part is easy (but the transitivity requires a simple game with absolute values).

The quotient space is with the indiscrete topology. Here is why: take any non-empty open subset U of the quotient space. So, there is $x \in U$. Denote the quotient mapping $\mathbb{R} \to \mathbb{R}/_{\sim}$ by p. Since U is open in $\mathbb{R}/_{\sim}$, $p^{-1}(U)$ is open in $\mathbb{R}$. We show that $p(p^{-1}(U)) = \mathbb{R}/_{\sim}$, and since $p(p^{-1}(U)) = U$, it would follow that $U = \mathbb{R}/_{\sim}$, so that $\mathbb{R}/_{\sim}$ is with the indiscrete topology. Since $p^{-1}(U)$ is open, it contains an interval I of finite length. Take any $[r] \in \mathbb{R}/_{\sim}$, where $r \in \mathbb{R}$. Then it suffices to find $x \in I$ with $r - x \in \mathbb{Q}$, which exists by density of rationals.

4.3 *The Gluing Lemma, Topological Sums, and Some Special Quotient Spaces*

Solutions of the odd-numbered exercises

---

1. (This exercise generalizes the Gluing Lemma.) Let X and Y be spaces, and let $f : X \to Y$ be a mapping.

(a) Let $X = \bigcup_{i \in I} U_i$, where each U_i is open in X, and suppose for every $i \in I$, the restriction $f|_{U_i}$ is continuous. Prove that f is continuous.

(b) Let $X = \bigcup_{i=1}^{n} F_i$, where each F_i is closed in X, and suppose for every $i \in \{1,2,...,n\}$, the restriction $f|_{F_i}$ is continuous. Prove that f is continuous.

(c) Does (b) remain true if we replace $\{F_i : i = 1,2,...,n\}$ by an infinite collection of closed subsets of X? Is (b) true if we do not assume that each F_i is closed?

Solution. (a) We will show that for open subset V of Y, $f^{-1}(V)$ is open in X. We observe:

$$f^{-1}(V) = f^{-1}(V) \cap X = f^{-1}(V) \cap \left(\bigcup_{i \in I} U_i \right) = \bigcup_{i \in I} \left(f^{-1}(V) \cap U_i \right). \text{ Since } f|_{U_i} \text{ is continuous,}$$

each $f^{-1}(V) \cap U_i$ is open in U_i. Since each U_i is open, $f^{-1}(V) \cap U_i$ is open in X. Since $f^{-1}(V)$ is a union of such sets, it is also open in X.

(b) Suffices to use the same argument as in part (a), replacing the words 'open' with the words 'closed' where they appear.

(c) No in both cases.

3. (a) Let $\{A_i : i = 1,2,3,...\}$ be a family of closed subsets of a metric space X such that $A_j \cap \left(\bigcup_{|i-j| \geq 2} A_i \right)' = \varnothing$, $j = 1,2,3,...$. such that $\bigcup_{i=1}^{\infty} A_i = X$, and such that $A_i \cap A_j = \varnothing$ for $|i-j| \geq 2$. Let $\{f_i : A_i \to Y : i = 1,2,3,...\}$ be a set of continuous mappings such that f_i and f_{i+1} agree on $A_i \cap A_{i+1}$, $i = 1,2,3,...$. Then the mapping $g : X \to Y$ defined by $g(x) = f_i(x)$ if $x \in A_i$ is well defined and continuous. [Hint: show that inverse images of closed sets are closed.]

(b) Show that the condition $A_j \cap \left(\bigcup_{|i-j| \geq 2} A_i \right)' = \varnothing$, $j = 1,2,3,...$ is necessary.

Solution. (a) That g is well defined follows from our assumptions (that f_i and f_{i+1} agree on $A_i \cap A_{i+1}$, $i = 1,2,3,...$, and that $A_i \cap A_j = \varnothing$ for $j \neq i+1$, $i = 1,2,3,...$.). So we only need to show that g is continuous. Take a closed subset F of Y. Then $g^{-1}(F) = \bigcup_{i=1}^{\infty} f_i^{-1}(F)$. If we show this is closed, we are done. Let a be an accumulation point of $g^{-1}(F)$. Since $\bigcup_{i=1}^{\infty} A_i = X$ this a is in some A_k. Since X is metric there is a sequence $\{x_j\}$ of elements of $g^{-1}(F)$ converging to a (Exercise 18 in 3.2). From some point on the members of that sequence must be in $\bigcup_{|k-i|<2} A_i$, else a would be in $\left(\bigcup_{|k-i|\geq 2} A_i\right)' \cap A_k = \varnothing$. So a is an accumulation point for the sequence in $\left(\bigcup_{|k-i|<2} A_i\right) \cap g^{-1}(F)$. Since $\left(\bigcup_{|k-i|<2} A_i\right) \cap g^{-1}(F) = \bigcup_{|k-i|<2} f_i^{-1}(F)$, this set is closed (as a union of not more than three closed sets). So $a \in \bigcup_{|k-i|<2} f_i^{-1}(F)$, and so $a \in g^{-1}(F)$. We proved that $g^{-1}(F)$ is closed.

(b) Let $X = \{0\} \cup \{\frac{1}{i+1} : i = 0,1,...\}$, be equipped with the subspace topology from $\mathbb{R}$, and take $A_i = \{\frac{1}{i+1}\}$, $i > 1$, $A_1 = \{0\}$, $f_i = identitity$ over A_i, $i > 1$ and $f_1(0) = 100$. Then the condition $\left(\bigcup_{|j-i|\geq 2} A_i\right)' \cap A_j = \varnothing$, $i = 1,2,3,...$ is not fulfilled and the conclusion in (a) fails.

5. Suppose $X_1 = \{0,a\}$ (a is not a number) is equipped with the indiscrete topology and suppose $X_2 = \mathbb{R}$. Describe the open subsets of the space X with the weak topology over $X_1 \cup X_2$ determined by $\{X_1, X_2\}$.

Solution. U is open in X if and only if it is one of the following two types:
  U is open in X_2 and $0 \notin U$,
  $U = V \cup \{a\}$, where V is open in X_2 containing 0.

7. Prove Proposition 2: A subset U of $\bigcup_{i \in I} X_i$ is open in $\bigoplus_{i \in I} X_i$ if and only if $U \cap X_i$ is open in X_i for every $i \in I$.

Solution. It follows directly from the definition of direct sums that U is open in X if and only if $U = \bigcup_{j \in J} U_j$, where each U_j is open in some X_i. After taking unions of those U_j-s that are in single X_i and re-indexing we may assume that $U = \bigcup_{i \in I} U_i$, where each U_i is open in X_i (we allow $U_i = \varnothing$ for some $i \in I$). The claim that U is such is clearly equivalent to the statement that $U \cap X_i$ is open in X_i for every $i \in I$.

9. Let $A = \{X_i : i \in I\}$ be a collection of (not necessarily disjoint) spaces and consider the set S of subsets of $\bigcup_{i \in I} X_i$ which are open in some X_i. Show that S is a subbasis for a topology τ over $\bigcup_{i \in I} X_i$, and that τ is finer than the weak topology over $\bigcup_{i \in I} X_i$ determined by A. Show that the inclusions $in_i : X_i \to \bigcup_{i \in I} X_i$ need not be continuous with respect to τ.

Solution. Since each X_i is open in itself, $X = \bigcup_{i \in I} X_i$ is a union of members of S, thus, by Proposition 3 in 3.3, S is a subbasis for a topology τ over X.

　　To show that τ is finer than the weak topology over X determined by A, we need to show that every set that is open with respect to the weak topology is also a member of τ. Open sets in the weak topology are of type $U \subset \bigcup_{i \in I} X_i$, where $U \cap X_i$ is open for every $i \in I$. Such sets clearly belong to τ.

　　Take $X_1 = \{1,2\}$ with the indiscrete topology, and take $X_2 = \{2\}$ with the singleton topology. Then $\{2\}$ is in τ, but $in_1^{-1}(\{2\}) = \{2\}$ is not open in X_1. Hence i_1 is not continuous.

11. Let X be the closed unit ball $\{(x,y,z) \in \mathbb{R}^3 : x^2 + y^2 + z^2 \le 1\}$. Denote $A = \{(x,y,z) \in X : x^2 + y^2 + z^2 = 1, z > 0\}$ and $B = \{(x,y,z) \in X : x^2 + y^2 + z^2 = 1, z < 0\}$. Define $f : A \to B$ by $f(x,y,z) = (x,y,-z)$. Show that X_f, the space obtained by identifying A and B along f, is homeomorphic to the three-sphere $S^3 = \{(x,y,z,u) \in \mathbb{R}^4 : x^2 + y^2 + z^2 + u^2 = 1\}$. (Compare with Example 7.)

Solution. Consider the two-sphere $S_u^2 = \{(x,y,z,u) \in S^3 : x^2 + y^2 + z^2 = 1 - u^2\}$, u is a fixed number in the interval $[-1,1]$. Then, obviously, $\bigcup_{u \in [-1,1]} S_u^2 = S^3$. Similarly, split X into subspaces $X_u = \{(x,y,z) \in X : x^2 + y^2 + z^2 \le 1, x = u\}$, u fixed in $[-1,1]$. Then, by

Example 6, after the identification along f each X_u becomes a two-sphere. Define

$g : X_u \to S_u^2$ by $g(u,v,z) = ((1 - u^2 - v^2)\cos\dfrac{2z\pi}{1 - u^2 - v^2}, (1 - u^2 - v^2)\sin\dfrac{2z\pi}{1 - u^2 - v^2}, v, u)$.

This factors nicely through the quotient space since $g(u,v,z) = g(u,v,-z)$ when $z = 1 - u^2 - v^2$ (i.e., when z is on the boundary of X.)

13. Let $f : A_1 \to A_2$ and $g : B_1 \to B_2$ be homeomorphisms. Let C be a subspace of A_1, let D be a subspace of B_1 and let $h : C \to D$ be a homeomorphism. Show that $A_1 \cup_h B_1$ is homeomorphic to $A_2 \cup_{(g|_D) \circ h \circ (f^{-1}|_{f(C)})} B_2$, the space we by identifying A_2 and B_2 along $(g|_D) \circ h \circ (f^{-1}|_{f(C)})$.

Solution (no details). The desired homeomorphism is $f \cup g$.

15. Let $A = X = [0,1]$, let $Y = [2,3]$ and let f be the mapping $A \to Y$ defined by

$f(x) = \begin{cases} 2 \text{ if } x < 1 \\ 3 \text{ if } x = 1 \end{cases}$. Show that the space $X \cup_f Y$ is not embeddable in any $\mathbb{R}^n$.

Solution. Since $\mathbb{R}^n$ is Hausdorff, so is any subspace. But $X \cup_f Y$ is not Hausdorff: Take an open set in $X \cup_f Y$ containing $[2]$. Now $[2] = [0,1)_X \cup \{2\}_Y$, where the index indicates the underlying space (as usual we use square brackets for equivalence classes). Every open subset U of $X \oplus Y$ that contains this set must be such that $U \supset [0,1)_X \cup [2,a)_Y$ for some $a \in (2,3]$. The corresponding open neighbourhoods of $[2]$ in $X \cup_f Y$ are the images of $[0,1)_X \cup [2,a)_Y$ under the quotient mapping. On the other hand $[3] = \{1\}_X \cup \{3\}_Y$, and open neighbourhoods of this set in $X \oplus Y$ must contain (as subsets) sets of type $(c,1]_X \cup (d,3]_Y$. Notice that $((c,1]_X \cup (d,3]_Y) \cap ([0,1)_X \cup [2,a)_Y) \neq \emptyset$. So, any two open sets around $[2]$ and $[3]$ respectively must intersect; so $X \cup_f Y$ is not Hausdorff.

17. Let X and Y be two spaces and let A be an open subspace of X and let $f : A \to Y$ be continuous. Show that the restriction of the quotient map $X \oplus Y \to X \cup_f Y$ over Y is an embedding.

Solution. For points in Y, we have $[y] = \{y\} \cup f^{-1}(y)$, so that y is the only point from y in this set. So, if two such classes of elements in Y are the same, then the corresponding elements in y are the same, and we are done.

4.4 Manifolds and CW-complexes

---

1. (a) Find a space X such that the conditions (i) and (iii) in the definition of the n-manifold are satisfied, and the condition (ii) fails.

(b) Find a space X such that the conditions (ii) and (iii) in the definition of the n-manifold are satisfied, and the condition (i) fails.

Solution. (a) Lexicographic order over $\mathbb{R}^2$ (Example 10 in 3.3) will do.

(b) Take two copies of the interval $[0,2]$ and identify each pair of corresponding points except the two 1-s.

3. A Möbius band is the quotient space obtained from a filled rectangle by identifying the two edges labeled a according to the arrows (left figure, Illustration 4.31). Start with two copies of a Möbius band and identify the boundaries according to the labels and arrows in the second and third figure in Illustration 4.31. Which 2-manifold is the resulting quotient space?

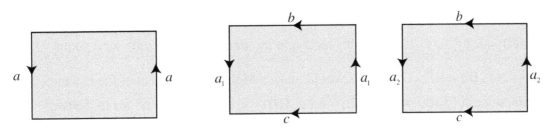

Illustration 4.31.

Solution: Klein bottle; to see that first identify the b and c edges (to get a cylinder), then the a-edges.

5. Let O be a subset of $\mathbb{R}^3$. Denote the rotation around the x-axis through α radians ($\alpha \in \mathbb{R}$) by f_α, and denote the rotation around the z-axis through γ radians ($\gamma \in \mathbb{R}$) by h_γ. The **configuration space** $C(O)$ of O with respect to the rotations f_α and h_γ (in that order) is the quotient space $\mathbb{R}^2\big/_{\sim}$ of $\mathbb{R}^2$, where $\sim$ is defined as follows: $(a,b) \sim (c,d)$ if $h_b(f_a(O)) = h_d(f_c(O))$. Describe the configuration space $C(O)$ if

      (a) $O = \{(0,0,z): -1 \leq z \leq 1\}$.

      (b) $O = \{(0,0,z): -1 \leq z \leq 1\} \cup \{(0,1,y): 0 \leq y \leq 0.5\} \cup \{(0,-1,y): 0 \leq y \leq 0.5\}$

(see Illustration 4.32).

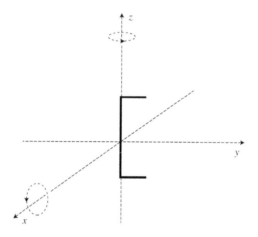

Illustration 4.32

Solution.

      (a) We achieve all positions of the object O with rotations around the x-axis through a radians, $0 \leq a \leq \frac{\pi}{2}$, and rotations around the z-axis through b radians, $0 \leq b \leq \pi$. This gives the rectangle $R = [0, \frac{\pi}{2}] \times [0, \pi] \subset \mathbb{R}^2$. Since $h_0(f_a(O)) = h_\pi(f_{\frac{\pi}{2}-a}(O))$, within this rectangle we identify points $(a,0)$ with points $(\frac{\pi}{2}-a, \pi)$, $0 \leq a \leq \frac{\pi}{2}$. Only this would make a Möbius band M out of the rectangle R. Further, since $h_b(f_0(O)) = h_d(f_0(O)) = h_e(f_{\frac{\pi}{2}}(O)) = h_f(f_{\frac{\pi}{2}}(O))$, the points in the two horizontal edges of R are all equivalent and so they are all identified to a single point. This means that the boundary the Möbius band M is shrunk into a single point. Hence, the space $\mathbb{R}^2\big/_{\sim}$ can be visualized as a Möbius band M with the boundary pinched down to a point, or a Klein bottle with one 'equatorial circle' – a circle obtained from the edges labeled a in Illustration 4.25 – contracted to a point.

      (b) The rectangle shown in the Figure 3 contains points in $\mathbb{R}^2$ that represent all equivalence classes of $\mathbb{R}^2\big/_{\sim}$. Different points in the interior of the rectangle represent different equivalence classes, and the points on the boundary are to be identified in $\mathbb{R}^2\big/_{\sim}$ as shown by the arrows and labels.

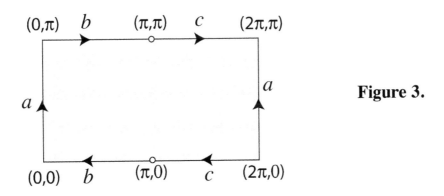

Figure 3.

(In Section 14.3 we present methods of recognizing 2-manifolds obtained from polygons by identifying pairs of edges. Using these methods it can be easily seen that the space $\mathbb{R}^2\!\big/_{\!\sim}$ in this case is a Klein bottle.)

7. Show that $\mathbb{R}^n$ is a CW-complex of dimension n.

Solution. $\mathbb{R}$ is a CW-complex of dimension 1 (integers are the 0skeleton, the line segments between them indicate how we attach the 1 cells. Assume $\mathbb{R}^n$ is an n-dimensional CW-complex realized as a mesh of n-dimensional cubes. Then $\mathbb{R}^{n+1} = \mathbb{R}^n \times \mathbb{R}$ is a $(n+1)$-dimensional CW-complex realized as a mesh of $(n+1)$-dimensional cubes. We attach the cells in the obvious way in all cases

9. [Adam Clay] Find a CW complex Y of dimension 2 such that Y is homeomorphic to $\mathbb{R}^2$ and no subcomplex of Y is homeomorphic to $\mathbb{R}$.

Solution [Clay] The 0-skeleton consists of all natural numbers. The 1-skeletons is indicated by the lines (circles and line segments) shown in Figure 4. The construction of the CW-complex Y is described in the same figure; the regions of different color/shade indicate how we attach 2-cells.

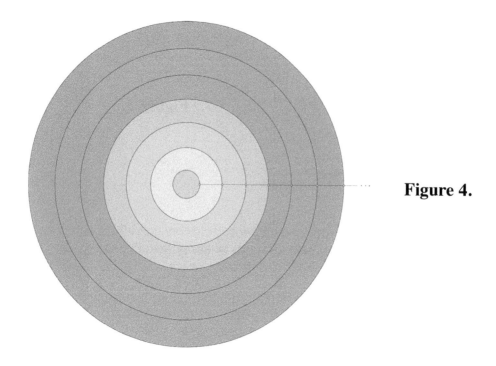

Figure 4.

11. Show that the k-skeleton X^k of a finite CW-complex is closed.

Solution. Suppose that X is a finite CW-complex of dimension $n+1$. It suffices to show that X^n is closed in X, since closed subsets of closed subsets of a space Y are also closed in Y. So, we prove that X^n is closed in $X = X^{n+1}$. Take a point in $X^{n+1} \setminus X^n$. So it comes from the interior of an $(n+1)$-cell (i.e., from the interior of an $(n+1)$-disk). Then the whole interior of this $(n+1)$-disk is an open set that is not disturbed in the identification process, hence it is open in $X = X^{n+1}$. We proved that $X^{n+1} \setminus X^n$ is open in X, hence X^n is closed in $X = X^{n+1}$.

13. Show that Klein bottle is a CW-complex of dimension 2.

Solution. Start with the 1-skeleton X^1 being the rose of two circles, labeled a and b. The attach the 2-cell shown in Illustration 4.25 so that the labels of the edges and their orientations match.

Chapter 5: Products of Spaces

Solutions of the odd-numbered exercises

---

1. Show that if $X_1 \cong X_2$ and $Y_1 \cong Y_2$, then $X_1 \times Y_1 \cong X_2 \times Y_2$.

Solution. If $f : X_1 \to X_2$ and $g : Y_1 \to Y_2$ are homeomorphisms, then so is $(f, g) : X_1 \times Y_1 \to X_2 \times Y_2$.

3. Show that if X is a discrete space, then the product space $X \times Y$ is homeomorphic to the discrete sum $\bigoplus_{x \in X} Y_x$, where each Y_x is a homeomorphic copy of the space Y.

Solution. Define $f : X \times Y \to \bigoplus_{x \in X} Y_x$ by $f(x,y) = y_x$, where y_x is the element in Y_x corresponding to y. It is plain that f is a bijection. Choose any $U \underset{open}{\subset} Y_x$. Then $f^{-1}(U) = \{x\} \times U$, and this is open in $X \times Y$. Since open subsets of Y_x-s (x changes) make a basis for $\bigoplus_{x \in X} Y_x$, it follows from Proposition 3 in 3.5, that f is continuous. Similarly, if $U \times V$ is a set from the standard basis of $X \times Y$, then $f(U \times V) = \bigoplus_{x \in U} V_x$, and this set is open in $\bigoplus_{x \in X} Y_x$. We proved that f is bijective, open and continuous; hence f is a homeomorphism.

5. Let A be a subspace of X and let B be a subspace of Y. Show that $A \times B$ with the product space topology is homeomorphic to $A \times B$ considered as a subspace of $X \times Y$.

Solution. It suffices to prove that every subset of $A \times B$ that is in one of the two topologies is also in the other. Suppose $U \times V$ is a set in the standard basis for $A \times B$. Since U is open in A, we have $U = A \cap U^*$ for some open subset U^* of X. Similarly, $V = A \cap V^*$ for some open subset V^* of Y. Then $U \times V = (A \times B) \cap (U^* \times V^*)$, and so $U \times V$ is open with respect to the subspace topology over $A \times B$. Since standard basis elements of $A \times B$ are open with respect to the subspace topology over $A \times B$, so are all open subset in the product space $A \times B$.

Conversely, suppose W is an open set in $A \times B$, considered as a subspace of $X \times Y$. So, $W = W^* \cap (A \times B)$ for some open subset W^* of $X \times Y$. We express W^* is a union of standard basis sets for the space $X \times Y$: $W^* = \bigcup_{i \in I}(U_i \times V_i)$. Then we compute:

$$W^* \cap (A \times B) = \left(\bigcup_{i \in I}(U_i \times V_i)\right) \cap (A \times B) = \bigcup_{i \in I}(U_i \times V_i) \cap (A \times B) = \bigcup_{i \in I}((U_i \cap A) \times (V_i \cap B)).$$

The last union shows that W is open in the product topology over $A \times B$.

7. Visualize the space $X \times Y$ if

(a) X is the number **8** (viewed as a subspace of $\mathbb{R}^2$) and Y is the (unit) circle S^1.

(b) The spaces X and Y are copies of the number **8** (viewed as a subspace of $\mathbb{R}^2$).

Solution. (a) Replace each point in by a circle, two distinct points giving rise to disjoint circles; the associated topology is as described in this section. Figure 5 is an imperfect rendering of the resulting space: as remarked, there should not be any intersection between the circles corresponding to different points.

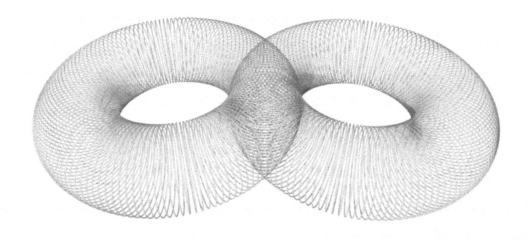

(b) Same comments as in (a) apply. A visualization of the space is given in Figure 6.

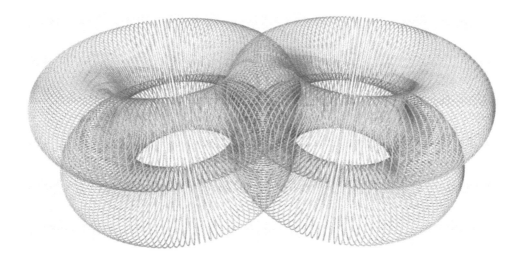

Figure 6.

9.   (a) Show that if each X_i, $i = 1, 2, \ldots, n$, is separable, then so is $\prod_{i=1}^{n} X_i$.

(b) Show that if each X_i, $i = 1, 2, \ldots, n$, is second countable, then so is $\prod_{i=1}^{n} X_i$.

Solution.

(a) Let D_i be a countable set that is dense in X_i, $i = 1, 2, \ldots, n$. Then $\prod_{i=1}^{n} D_i$ is countable (Proposition 4, Section 1.2). The simplest at this point is to use Exercise 13 (solution below): $\overline{\prod_{i=1}^{n} D_i} = \prod_{i=1}^{n} \overline{D_i} = \prod_{i=1}^{n} X_i$.

(b) We prove the claim for $n = 2$, and the rest follows by induction. Let $\mathcal{B}$ be a countable basis for X and let $\mathcal{C}$ be a countable basis for Y. By Proposition 4, Section 1.2, $\mathcal{B} \times \mathcal{C}$ is countable. Let U be open in X and V be open in Y. Then $U = \bigcup_{i \in I} B_i$ for some $B_i \in \mathcal{B}$, and $V = \bigcup_{j \in J} C_j$ for some $C_j \in \mathcal{C}$. Then $U \times V = \left(\bigcup_{i \in I} B_i \right) \times \left(\bigcup_{j \in J} C_j \right) = \bigcup_{\substack{i \in I \\ j \in J}} \left(B_i \times C_j \right)$.

With this we managed to express every set in the standard basis for $X \times Y$ as a union of elements of $\mathcal{B} \times \mathcal{C}$. Consequently, every open set in $X \times Y$ is a union of elements of $\mathcal{B} \times \mathcal{C}$, showing that $\mathcal{B} \times \mathcal{C}$ is a basis for $X \times Y$.

11. Let $f : X \rightarrow Y$ be continuous, and let Z be the subspace of $X \times Y$ defined as follows: $Z = \{(x, y) \in X \times Y : y = f(x)\}$.

(a) Show that Z is homeomorphic to X.

(b) Show that if Y is Hausdorff, then Z is closed in $X \times Y$.

Solution.

(a) Define $g : X \to Z$ by $g(x) = (x, f(x))$. It is clear that g is continuous and bijective. To see it is open take and open subset U of X. Then $f(U) = \{(x, f(x)) : x \in U\}$. Now observe that $\{(x, f(x)) : x \in U\} = (U \times Y) \cap Z$ and the story is over.

(b) Suppose $(x, y) \notin Z$. So, $y \neq f(x)$. Since Y is Hausdorff there are disjoint open U_y and $U_{f(x)}$. Since f is continuous, there is U_x such that $f(U_x) \subset U_{f(x)}$. So, $U_x \times U_y$ is an open set containing (x, y) that is disjoint from Z. Otherwise, if $(a, f(a)) \in U_x \times U_y$, then $f(a)$ also belongs to $U_{f(x)}$, and this last set was chosen to be disjoint from U_y.

13. Let A_i be a subset of the space X_i, $i = 1, 2, \ldots, n$ (so that $\prod_{i=1}^{n} A_i$ is a subset of $\prod_{i=1}^{n} X_i$). Show that $\overline{\prod_{i=1}^{n} A_i} = \prod_{i=1}^{n} \overline{A_i}$. (Hint: do it for $n = 2$ and then use induction.)

Solution. Following the hint, we prove that $\overline{A \times B} = \overline{A} \times \overline{B}$ for spaces X and Y and sets $A \subset X$ and $B \subset Y$. Since $\overline{A} \times \overline{B}$ is closed and $A \times B \subset \overline{A} \times \overline{B}$, it follows that $\overline{A \times B} \subset \overline{A} \times \overline{B}$. For the converse inclusion, choose $(a, b) \in \overline{A} \times \overline{B}$. For every open neighborhood W of (a, b) there is a standard basis set $U \times V$ such that $(a, b) \in U \times V \subset W$. Since $U \cap A \neq \varnothing$ and $V \cap B \neq \varnothing$, it follows that $(U \times V) \cap (A \times B) \neq \varnothing$. Hence $W \cap (A \times B) \neq \varnothing$ for every open neighborhood W of (a, b). This proves that $(a, b) \in \overline{A \times B}$, hence $\overline{A} \times \overline{B} \subset \overline{A \times B}$, and so $\overline{A \times B} = \overline{A} \times \overline{B}$.

15. Show that if Y is Hausdorff and $f : X \to Y$ is continuous, then $\{(x_1, x_2) \in X \times X : f(x_1) = f(x_2)\}$ is a closed subset of $X \times X$.

Solution. Denote $A = \{(x_1, x_2) \in X \times X : f(x_1) = f(x_2)\}$. We show that the complement A^c is open. Choose any $(u, v) \in A^c$, so that $f(u) \neq f(v)$. Since Y is Hausdorff, there are open neighborhoods U and V of $f(u)$ and $f(v)$ respectively, such that $U \cap V = \varnothing$. Then $f^{-1}(U)$ and $f^{-1}(V)$ are open and disjoint neighborhoods of u and v respectively. Hence $f^{-1}(U) \times f^{-1}(V)$ is an open neighborhood of (u, v) in $X \times X$. We show that $f^{-1}(U) \times f^{-1}(V) \subset A^c$, establishing that A^c is open, as claimed. Suppose $f^{-1}(U) \times f^{-1}(V)$ is not a subset of A^c. Then there are $(s, t) \in f^{-1}(U) \times f^{-1}(V)$ such that $f(s) = f(t)$. However, since $f(s) \in U$ and $f(t) \in V$, this would imply that $U \cap V \neq \varnothing$, which establishes a contradiction.

17. Describe the 3-manifold introduced in Example 7, 4.4, as a product of known topological spaces.

Solution. It is (obviously) $S^2 \times S^1$.

19. (a) Find a subset U of $\mathbb{R}^2$ such that the intersection of U with any horizontal or vertical line is open in that line (considered as a subspace of $\mathbb{R}^2$), yet U is not open in $\mathbb{R}^2$.

(b) Show that the usual topology over $X \times Y$ is, in general, strictly weaker than the topology induced by all coordinate mappings $c_1^{z_0} : X \to X \times Y$, $c_2^{z_0} : Y \to X \times Y$, where $z_0 = (x_0, y_0)$ ranges through $X \times Y$ and where $c_1^{z_0}(x) = (x, y_0)$, and $c_2^{z_0}(y) = (x_0, y)$.

Solution.

(a) Take F to be a sequence converging to $(1,1)$ (without $(1,1)$); then F is not closed, so that $U = F^c$ is not open. It can now be readily seen that U intersection any vertical or horizontal line is open. (Simpler, take y=-x without the origin.)

(b) Denote by $\mathcal{T}_1$ the product topology; let $\mathcal{T}_2$ stand for the topology induced by all $c_1^{z_0}$, $c_2^{z_0}$, where $z_0 = (x_0, y_0)$ ranges through all elements of $X \times Y$. By Proposition 7 of 3.5, we have that

$$\mathcal{T}_2 = \left\{ Z \subset X \times Y : \left(c_1^{z_0}\right)^{-1}(Z) \underset{open}{\subseteq} X \text{ and } \left(c_2^{z_0}\right)^{-1}(Z) \underset{open}{\subseteq} Y, \text{ for all } z_0 \in X \times Y \right\}. \text{ Each}$$

standard basis element $U \times V$ for $\mathcal{T}_1$ is in $\mathcal{T}_2$. That is true since $\left(c_i^{z_0}\right)^{-1}(U \times V)$ is an open set (open in X for $i = 1$, open in Y for $i = 2$). For example, $\left(c_1^{z_0}\right)^{-1}(U \times V) = U$ if $y_0 \in V$, $\left(c_1^{z_0}\right)^{-1}(U \times V) = \varnothing$ if $y_0 \notin V$ (a similar statement holds for $c_2^{z_0}$). Consequently $\mathcal{T}_1 \subset \mathcal{T}_2$. That the inclusion can be strict is shown in (a).

Alternative solution to (b): The product topology is the smallest topology that makes the projection maps continuous. So it suffices to check that with the induced topology from the coordinate mappings, that the projection maps are continuous. But with this topology, continuous functions are precisely those which when precomposed with the coordinate mappings (so that the coordinate mapping acts first) are continuous. For the projection maps, the only possibilities after precomposing are constant functions and identities, and we're done.

5.2 Infinite Products of Spaces

Solutions of the odd-numbered exercises

---

1. Show that each projection $p_j : \prod_{i \in I} X_i \to X_j$ is open.

Solution. Consider first the standard basis set $\prod_{i \in I} U_i$, where all U_i-s are open and where in all but finitely many cases $U_i = X_i$. Then $p_j \left(\prod_{i \in I} U_i \right) = U_j$, and this is open. Since every open set is union of basis sets, and since for every mapping $f \left(\bigcup_{i \in I} Z_i \right) = \bigcup_{i \in I} f(Z_i)$, the claim in this exercise follows.

3. Show that it is not true that if each of the spaces X_i is separable (first countable, second countable), then the space $\prod_{i \in I} X_i$ must be separable (first countable, second countable, respectively).

Solution. $\mathbb{R}$ is separable, first countable and second countable, and it is clear that $\prod_{i \in \mathbb{R}} \mathbb{R}$ is neither first countable nor second countable.

However, and somewhat surprisingly, this space is separable. (Exercise: prove it!] In order for $\prod_{i \in I} \mathbb{R}$ not to be separable it is sufficient and necessary that the index set I be of cardinality larger than $2^{\aleph_0} = |\mathbb{R}|$. Here is a proof that $\prod_{i \in I} \mathbb{R}$ is not separable if $|I| > |\mathbb{R}|$.

Denote $X = \prod_{i \in I} \mathbb{R}$, and assume it is separable; we prove that in that case $|I| \le |\mathbb{R}|$.

Under our assumption there is a countable dense set D. Denote by p_i the projection $X \to \mathbb{R}$ onto the i-th component of the product space, and let U be any (non-empty) open interval $(a,b) \subsetneq \mathbb{R}$. Since D is dense, each subset $D_i = p_i^{-1}(U) \cap D$ of D, $i \in I$, is not empty. Now we show that $i \mapsto D_i$ is a one-to-one mapping $I \to \mathcal{P}(D)$, where $\mathcal{P}(D)$ is the power set of D. It would follow from the definition of the relation $<$ among cardinal numbers that $|I| \le |\mathcal{P}(D)| = 2^{\aleph_0}$, which is what we wanted to show.

Choose any $i, j \in I$, $i \ne j$, and let V be any non-empty open subset of $\mathbb{R}$ that is disjoint from U. Then $p_i^{-1}(U) \cap p_j^{-1}(V)$ is open and non-empty, and so, since D is dense, there is a point x in $p_i^{-1}(U) \cap p_j^{-1}(V) \cap D$. Now, this x is not in $p_j^{-1}(U) \cap D$, since U and V are disjoint. Hence $x \in p_i^{-1}(U) \cap p_j^{-1}(V) \cap D \subset D_i$, and $x \notin p_j^{-1}(U) \cap D = D_j$, implying that $D_i \neq D_j$. This proves that $i \mapsto D_i$ is one-to-one.

5. Show that if I is infinite, and if for every $i \in I$, X_i is homeomorphic to a fixed space X, then for any fixed $j \in I$, the spaces $\prod_{i \in I} X_i$ and $\prod_{\substack{i \in I \\ i \neq j}} X_i$ are homeomorphic.

Solution. There is a bijection $f : I \to I \setminus \{j\}$; to find it isolate a countable subset C containing j, find a bijection $C \to C \setminus \{j\}$ and fix the elements in $I \setminus C$. Now define

$$g : \prod_{i \in I} X_i \to \prod_{\substack{i \in I \\ i \neq j}} X_i \text{ as follows: } g\big((x_i)_{i \in I}\big) = (x_{f(i)})_{i \in I} \text{, where for } x \in X, \ x_k \text{ is the}$$

corresponding element in X_k. It is now straightforward to show that g is a homeomorphism.

7. Show that if I is infinite then the set $\prod_{i \in I} U_i$, where each U_i is a non-empty proper open subset of X_i, is NOT open in $\prod_{i \in I} X_i$.

Solution. Of course. None of the standard basis elements is contained in $\prod_{i \in I} U_i$ so it could not be a union of such.

9. (a) Let $X = \mathbb{N}$ be equipped with any topology. Show that X^∞ is separable.

(b) Show that if X_i, $i = 1, 2, \ldots, n, \ldots$, are spaces such that $|X_i| = \aleph_0$, then $\prod_{i=1}^{\infty} X_i$ is separable

Solution. Remark: note that $\prod_{i=1}^{\infty} X_i$ need not be countable, so the problem is not entirely trivial.

We prove (a); similar argument applies to (b).

Fix $a \in X$. There are countably many elements of type $(x_1, a, a, \ldots, a, \ldots)$ in X^∞. So, since finite products of countable sets are countable, there are countably many elements of the space X^∞ of type $(x_1, x_2, x_3, \ldots, x_n, a, a, \ldots)$, Take the unions of all such: countable union of countable sets is countable. Get a countable set D of all sequences in X^∞ that eventually stabilize at a. We claim that D is dense in X^∞. Suffices to show that every open set in the standard basis of X^∞ has a non-empty intersection with D. But every such standard basis element is of the type $\prod\limits_{i=1}^{\infty} U_i$ where there is a natural number N such that for $n > N$ all U_i are X. So, the element $(x_1, x_2, x_3, ..., x_N, a, a, ...)$ with $x_i \in U_i$, $i = 1, 2, ..., N$ is an element from D that is also in $\prod\limits_{i=1}^{\infty} U_i$.

11. Let $\bar{\mathbf{x}} = (x_i)_{i \in I}$ be a fixed element in $\prod\limits_{i \in I} X_i$. For every finite subset T of I, denote $C(T) = \prod\limits_{i \in I} A_i$, where $A_i = \{x_i\}$ if $i \notin T$, and $A_i = X_i$ if $i \in T$. Show that the set $Y = \bigcup\limits_{\substack{T \subseteq I \\ finite}} C(T)$ is dense in $\prod\limits_{i \in I} X_i$.

Solution. Take any standard basis set $\prod\limits_{i \in I} U_i$ for the product space $\prod\limits_{i \in I} X_i$. This means that all U_i-s are open and non-empty and in only finitely many cases $U_i \neq X_i$. Suppose the latter happens for $T = \{i_1, i_2, ..., i_n\} \subset I$. Then $C(T) \cap \prod\limits_{i \in I} U_i \neq \varnothing$, hence $Y \cap \prod\limits_{i \in I} U_i \neq \varnothing$. It follows that $Y \cap W \neq \varnothing$ for every open subset W of $\prod\limits_{i \in I} X_i$.

13. Show that the mapping $f(0. x_1 x_2 \cdots x_n \cdots) = \prod\limits_{i=1}^{\infty} x_i$ from the Cantor set C to product space $\prod\limits_{i \in I} X_i$ (with each X_i a copy of the discrete space over $\{0,2\}$, as in Example 5) is indeed a homeomorphism.

Solution. It is clear that f is a bijection.

Since f is a bijection, in order to prove it is continuous it suffices to show that the inverse images of sub-basis sets of type $\{0,2\} \times \{0,2\} \times \cdots \times \{0,2\} \times \{p\} \times \{0,2\} \times \cdots$, $p \in \{0,2\}$ (and this p appears as the n-th coordinate) are open in C. That these sets are open is evident since any of them is a union of sets of elements in C belonging to intervals I_{n_m} in the sets C_n as described in the construction of C in the text, and since $I_{n_m} \cap C$ is open for every n and m.

[Before we prove that f is open consider an example: let's describe all numbers in C that are between $a=0.020020002...$ and $b=0.2020202020...$. The first place these two differ is at the first decimal. If we keep the 0 in the first decimal of a, then any time we have a 0 in a we can change the 0 to 2 and get a number between a and b. So, using xxx to denote 0 or 2 at a decimal place, we have the following sets: 0.022xxx, then 0.0200202xxx, etc.

The mapping f sends these to standard basis open sets in the product space $\prod_{i \in I} X_i$, and so their union is also open. In case we change the first decimal digit in a form 0 to 2, then we need to see if a is still less than b. It is, and then we go to the first instance we have different digits at a fixed decimal place of a and b. There we must have 0 in case of a and 2 in case of b. We keep these two and then increase each case of 0 in a fixed decimal place of a to 2 and repeat the procedure in the first case. The conclusion is the same, and f sends the corresponding sets to basic open sets in $\prod_{i \in I} X_i$. Iterating this procedure we cover the whole set of numbers between a and b in the Cantor set C, and so its image under f is open. On the side of b we proceed symmetrically. Keeping the digits in the first decimal place unchanged, we look again for the first instance we have a 0 in a and a 2 in b, but this time we decrease the later. Thus we get 0.200xxx. In the next step we get 0.202020xxx etc. Again we get sets that are clearly mapped to open (standard basis) sets in the product space $\prod_{i \in I} X_i$. Finally, we must see if the first decimal digits in b can be decreased, still getting a number larger than a. This does not happen in this case, since 0.0020202020... is less than a. So we have exhausted all possibilities for elements in C between a and b, and we saw along the way that f sends the whole set to unions of open sets in $\prod_{i \in I} X_i$.]

Now we tackle the general event and prove that f is open: we take the set of all Cantor numbers between two fixed $a, b \in C$, $a < b$. We focus on the case a and b are included in our set (when this assumption does not contradicts specifics of the cases we cover): since singletons are closed in both C and $\prod_{i \in I} X_i$, the other cases are easy consequences. Note that such type of open sets are intersections of open intervals and C, and so they are a basis for C. Proving that they are sent to open sets would then establish that f is open. We now describe the set $[a,b] \cap C$ as a union of open sets. Denote $a = 0.n_1 n_2 n_3 ...$, and $b = 0.m_1 m_2 m_3 ...$. Suppose the first time these decimal digits differ as when the index is k, so that $n_k = 0$ and $m_k = 2$. Now we search for $l > k$ such that $n_l = 0$. If this happens then we change n_l to 2 and choose any digits 0 or 2 after this decimal place; the numbers $0.n_1 n_2 n_3 ... n_{l-1} 2 xxx$ (where the part xxx denotes any choice of 0 or 2 in the decimal places following the l-th) will be between a and b. Then look at a again and look for a decimal place $l_2 > l$ where $n_{l_2} = 0$. If this happens then we change n_{l_2} to 2, and let the remaining decimals be any digit 0 or 2; we have the numbers $0.n_1 n_2 n_3 ... n_{l_2 - 1} 2 xxx$ (same meaning of xxx) will be between a and b. Keep doing this; all of these sets are clearly mapped to open (standard basis) sets in $\prod_{i \in I} X_i$. For example, $0.n_1 n_2 n_3 ... n_{l-1} 2 xxx$ goes to

$$\{n_1\} \times \{n_2\} \times \cdots \times \{n_l\} \times \{2\} \times X \times X \times \cdots.$$

Now, change n_k from 0 to 2 to get the decimal $0.n_1 n_2 n_3 ... n_{k-1} 2 n_{k+1} ...$. If this is larger than b then we stop here; otherwise you search for the first decimal $s > k$ where $n_s = 0$ and

$m_s = 2$. Change the s to k (notation-wise) and repeat the procedure in the preceding paragraph. Keep going. The images under f of the sets we get along the way are open (standard basis) sets in the product space $\prod_{i \in I} X_i$, and hence so is the image of the union of these sets.

Now we deal with b. We note in passing that what we do below covers the possibility when n_{l_1}, n_{l_2}, ... do not exist. For example when there is no n_l as above, then $a = 0.n_1 n_2 .. n_{k-1} 0222222...$ and $b = 0.n_1 n_2 .. n_{k-1} 222222...$; what we do below covers this case too.

So, focusing on b, we do what we did to a interchanging the roles of 0 and 2. SO, just to start this off, suppose the first time these decimal digits differ as when the index is k, so that $n_k = 0$ and $m_k = 2$. Now we search for $l > k$ such that $m_l = 2$. If this happens then we change m_l to 0 and choose any digits 0 or 2 after this decimal place; the numbers $0.m_1 m_2 .. m_{l-1} 0xxx$ (where the part xxx denotes any choice of 0 or 2 in the decimal places following the l-th) will all be between a and b. The argument used for a applies mutatis mutandis.

Note: A simpler solutions relies on the theory developed in Section 7.2, where we prove that a continuous bijection from a *compact* space onto a Hausdorff space must be a homeomorphism.

15.   (a) Show that the Cantor set C, considered as a subspace of $\mathbb{R}$, is a fractal.
(b) Visualize the product spaces $C \times C$ and $C \times C \times C$.
(c) Show that if X is a fractal then the product space X^n is a fractal.

Solution.

(a) $f : C \to C$ defined by $f(x) = \dfrac{1}{3}x$ is a contraction that sends C into a proper subset of C.

(b) Figures 7 and 8: $C \times C$, then $C \times C \times C$; 4 iterations in both cases (POV-Ray)

Figure 7.

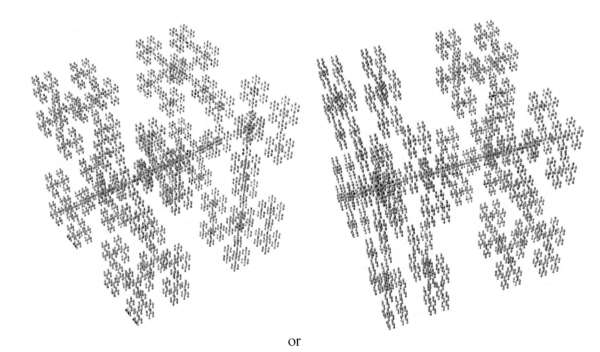

or

Figure 8.

(c) Suppose $f : X \to X$ is a contraction ($i = 1, 2, ..., n$). This means that for every $x, y \in X$, $d(f(x), f(y)) = \alpha d(x, y)$, for some $0 < \alpha < 1$. Define $g : X^n \to X^n$ by $g(x_1, x_2, ..., x_n) = (f(x_1), f(x_2), ..., f(x_n))$. In order to check this is a contraction suffices to show that

$$\sqrt{d(f(x_1), f(y_1))^2 + d(f(x_2), f(y_2))^2 + ... + d(f(x_n), f(y_n))^2} =$$
$$= \alpha \sqrt{d(x_1, y_1)^2 + d(x_2, y_2)^2 + ... + d(x_n, y_n)^2}.$$

That, however, is very easy.

5.3 Box Topology

Solutions of odd-numbered exercises

---

1. For a collection X_i, $i \in I$ of spaces, show that the set of all sets of type $\prod_{i \in I} U_i$, U_i is open in X_i, is a basis for a topology over the set $\prod_{i \in I} X_i$.

Solution. Since $\prod_{i \in I} X_i$ is itself a set of the given type, we only need to show that for every $B_1 = \prod_{i \in I} U_i$ and $B_2 = \prod_{i \in I} V_i$ ($U_i - s$, $V_i - s$ open in the corresponding $X_i - s$) and every $x \in B_1 \cap B_2$, there is a $B_3 = \prod_{i \in I} W_i$, $W_i - s$ open, such that $x \in B_3 \subset B_1 \cap B_2$. This is true since we can take $B_3 = \prod_{i \in I} (U_i \cap V_i)$.

3. Show that $\overline{\prod_{i \in I}^{Box} A_i} = \prod_{i \in I}^{Box} \overline{A_i}$.

Solution. The set $\prod_{i \in I}^{Box} \overline{A_i}$ is closed (in the box topology) since its complement is the open set $\bigcup_{i \in I} \left(\prod_{\substack{j \in I \\ i \neq j}} X_j \right) \times V_i$, V_i open in X_i. Since $\overline{\prod_{i \in I}^{Box} A_i}$ is the smallest closed set containing $\prod_{i \in I}^{Box} A_i$, it follows that $\overline{\prod_{i \in I}^{Box} A_i} \subset \prod_{i \in I}^{Box} \overline{A_i}$. In order to prove the opposite inclusion, take an element $\mathbf{x} \in \prod_{i \in I}^{Box} \overline{A_i}$. If $\mathbf{x} \in \prod_{i \in I}^{Box} A_i$, then clearly $\mathbf{x} \in \overline{\prod_{i \in I}^{Box} A_i}$. Suppose otherwise. Then $\mathbf{x} = (x_i)$, and for at least one j, $x_j \in A_j' \setminus A_j$. Take any standard basis open set $\prod_{i \in I}^{Box} V_i$ (for the box topology) containing $\mathbf{x}$. Then $\left(\left(\prod_{i \in I}^{Box} V_i \right) \setminus \{\mathbf{x}\} \right) \cap \left(\prod_{i \in I}^{Box} A_i \right) \neq \varnothing$, and so $\mathbf{x} \in \left(\prod_{i \in I}^{Box} A_i \right)'$. Consequently $\mathbf{x} \in \overline{\prod_{i \in I}^{Box} A_i}$. We proved that $\prod_{i \in I}^{Box} \overline{A_i} \subset \overline{\prod_{i \in I}^{Box} A_i}$.

5 [revised]. Consider the set $X = \prod_{i\in\mathbb{R}} \mathbb{R}$ of all functions $\mathbb{R} \to \mathbb{R}$; let B the set of all

bounded functions in X; so $f \in B$ if there are numbers $u,v \in \mathbb{R}$, such that $u \le f(x) \le v$
for every $x \in \mathbb{R}$.

     (a) Show that the sets of type $\prod_{i\in\mathbb{R}}(f(i)-\varepsilon(i), f(i)+\varepsilon(i))$, $f, \epsilon \in B$, $\varepsilon(i) > 0$,

comprise a basis for a topology $\mathcal{T}$ over B. [Here $(f(i)-\varepsilon(i), f(i)+\varepsilon(i))$ denotes an
interval; two simple basis sets are shown in Figure 2 (the functions that play the roles of
ϵ are both constants in the illustration).]

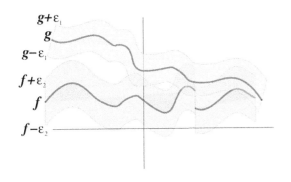

Illustration 5.15. Two sets from our basis of bounded real valued functions. What is their intersection? Careful!

     (b) Show that the topology $\mathcal{T}$ is the subspace topology of the box topology over
$X = \prod_{i\in\mathbb{R}} \mathbb{R}$.

     (c) Define $\rho : B \times B \to \mathbb{R}$ by $\rho(f,g) = \sup\{|f(t)-g(t)| : t \in \mathbb{R}\}$. Show that ρ is a
metric.

     (d) Show that the topology $\mathcal{T}$ is finer than the topology of the metric space
(B,ρ).

Solution.

     (a) It is obvious that B is the union of all $\prod_{i\in\mathbb{R}}(f(i)-\varepsilon(i), f(i)+\varepsilon(i))$. So, we need

to show that for every $\nabla \in \left(\prod_{i\in\mathbb{R}}(f(i)-\varepsilon(i), f(i)+\varepsilon(i))\right) \cap \left(\prod_{i\in\mathbb{R}}(g(i)-\delta(i), g(i)+\delta(i))\right)$, (

$f, g, \epsilon, \delta \in B$, $\epsilon, \delta > 0$), there exists $\prod_{i\in\mathbb{R}}(h(i)-\gamma(i), h(i)+\gamma(i))$, ($h, \gamma \in B, \gamma > 0$) such that

$\nabla \in \prod_{i\in\mathbb{R}}(h(i)-\gamma(i), h(i)+\gamma(i)) \subset \left(\prod_{i\in\mathbb{R}}(f(i)-\varepsilon(i), f(i)+\varepsilon(i))\right) \cap \left(\prod_{i\in\mathbb{R}}(g(i)-\delta(i), g(i)+\delta(i))\right)$.

This is easy: we can define $h(i)$ to be the middle of the open interval

$(f(i)-\varepsilon(i), f(i)+\varepsilon(i))\cap(g(i)-\delta(i), g(i)+\delta(i))$, which is not empty since $\nabla(i)$ is in it, and we can define $\gamma(i)$ to be any number smaller than one half of the interval $(f(i)-\varepsilon(i), f(i)+\varepsilon(i))\cap(g(i)-\delta(i), g(i)+\delta(i))$.

(b) It is obvious that B is a subset of $X=\prod_{i\in\mathbb{R}}\mathbb{R}$. It is also obvious that every open set in B (with topology $\mathcal{T}$ as defined in part (a)) is also open in the box topology over X. Consequently, every open set U in B is the intersection of B with the open set U in X.

(c) The Least Upper Bound property guarantees that $\rho(f,g)$ exists. The only property that is not entirely obvious is the triangular property $\rho(f,g)\le\rho(f,h)+\rho(h,g)$. This translates into the following inequality:
$\sup\{|f(t)-g(t)|: t\in\mathbb{R}\}\le\sup\{|f(t)-h(t)|: t\in\mathbb{R}\}+\sup\{|h(t)-g(t)|: t\in\mathbb{R}\}$. This in turn follows from $|f(t)-g(t)|=|f(t)-h(t)+h(t)-g(t)|\le|f(t)-h(t)|+|h(t)-g(t)|$.

(d) Every interval in the metric space (B,ρ) is of the type
$\prod_{i\in\mathbb{R}}(f(i)-c, f(i)+c)$, where $f\in B$ and $c>0$. It is clear that this set is in $\mathcal{T}$. Since the intervals of this type are a basis for the metric space (B,ρ), it follows that every open set in (B,ρ) is also in $\mathcal{T}$.

Chapter 6: Connected Spaces and Path Connected Spaces

6.1 Connected Spaces: Definition and Basic Facts

Solutions of the odd-numbered exercises.

---

1. Show that X is disconnected if and only if there are two non-empty open subsets A and B of X, such that $X = A \cup B$ and $\bar{A} \cap B = \varnothing = A \cap \bar{B}$.

Solution. $\Leftarrow$ is obvious.

$\Rightarrow$: Suppose X is disconnected; hence there are two non-empty open subsets A and B of X, such that $X = A \cup B$ and $A \cap B = \varnothing = A \cap B$. We prove that $\bar{A} \cap B = \varnothing$; it then follows by symmetry that $\varnothing = A \cap \bar{B}$. Assume for a moment that $\bar{A} \cap B \neq \varnothing$. Since we have assumed earlier that $A \cap B = \varnothing$, there must be $a \in (A' \setminus A) \cap B$. But then B is an open neighborhood of a such that $A \cap B = \varnothing$, contradicting the definition of A'. Hence $\bar{A} \cap B = \varnothing$.

3. Let A_i, $i = 1,2,\dots$ be a collection of connected subsets of a space X such that $A_i \cap A_{i+1} \neq \varnothing$ for every $i = 1,2,\dots$. Show that $\bigcup_{i=1}^{\infty} A_i$ is connected.

Solution. Assume that $\bigcup_{i=1}^{\infty} A_i$ is disconnected, i.e., assume that $\bigcup_{i=1}^{\infty} A_i = B \cup C$ for two open, non-empty, disjoint sets B and C. If for some $i \in \mathbb{Z}^+$, $\varnothing \neq A_i \cap B \neq A_i$, then $A_i \cap B$, $A_i \cap C$ is a disconnection of A_i, contradicting the setup of this question. Hence $B = \bigcup_{j \in J} A_j$, and $C = \bigcup_{k \in K} A_k$ for some two nonempty, disjoint subsets of $\mathbb{Z}^+$, such that $\mathbb{Z}^+ = J \cup K$. We may suppose that $1 \in J$. Since K is not empty, there is $n \in \mathbb{Z}^+$ such that $n \in J$ and $n+1 \in K$. But then $A_n \cap A_{n+1} = \varnothing$, getting a contradiction again. Hence $\bigcup_{i=1}^{\infty} A_i$ must be connected.

5. Show that for every x in a space X, the component of X containing x is the union of all connected subsets of X containing x.

Solution. Suppose C is the component of X containing x. If C_x is a connected set containing x, then $C_x \subset C$ or else $C_x \cup C$ would be larger than C and connected (Proposition 6(a)). It follows that the union of all connected subsets of X containing x is a subset of C. However, it is obvious that C is a subset of the union of all connected subsets of X containing x, since C is one of the members of the union!

7. Find three distinct connected subsets A, B, and C of $\mathbb{R}^2$ such that $\partial A = \partial B = \partial C$.

Solution. Take A to be the upper open half-plane plus the origin, take B to be the lower half plane, take C to be the union of A and B.

9.   Describe the components of the space X.
        (a) X is the subspace of $\mathbb{R}$ consisting of all irrational numbers.
        (b) X is the half-open interval topology over $\mathbb{R}$.

Answer: (a) Singletons. (b) Singletons.

11. Find a subspace of $\mathbb{R}^2$ with uncountably many components, all of them uncountable.

Solution. Take the subspace of $\mathbb{R}^2$ made of all vertical lines passing through the irrational numbers of the x-axis.

13. Show that every countable set A in any metric space such that $|A| > 1$ is disconnected.

Solution. Since A is countable, $A = \{a_1, a_2, ...\}$ (finite or infinite). Denote $d_i = d(a_1, a_i)$, $i > 1$ and where d is the metric. Since A is countable there is $r > 0$ such that $r \neq d_i$ for every $i > 1$, and such that $r < d_2$. Then $B(a_1, r) \cap A$ and $\left(\overline{B(a_1, r)} \right)^c \cap A$, where the complement is taken in the metric space, is a disconnection of A.

15. Show that no closed interval in $\mathbb{R}$ is a union of more than one and countably many pairwise disjoint closed sets.

Solution.

Without any loss of generality we may suppose that the closed interval is the unit interval $I = [0,1]$. The set I is not a finite non-trivial (more than one member) union of pairwise disjoint closed sets, since otherwise I would be disconnected, contradicting Proposition 2.

Suppose $I = \bigcup_{i=1}^{\infty} F_i$, where each F_i is a closed subset of I, and where $F_i \cap F_j = \varnothing$ for $i \neq j$. We construct a countable list of nested open intervals as follows. We may suppose that $\{0,1\} \subset F_1$ (just take F_1 to be the union of ≤ 2 sets containing 0 and 1, then re-index if needed). Since F_1^c is not empty and open in I, there is a maximal open interval $O_1 = (a_1,b_1)$ that is a subset of F_1^c (it is maximal in the sense that no other interval in F_1^c is an overset of O_1; consequently, both a_1 and b_1 are in F_1).

[This paragraph will be partially eclipsed by the next.] Since F_2 is closed and disjoint from F_1 both a_1 and b_1 are not in F_2. (Similar argument implies that a_1 and b_1 are not in F_i for all $i \geq 2$.) Focus on b_1: it is in F_2^c, and since the latter is open, there is a maximal open interval (a_2, b_2') containing b_1 and within F_2^c. Set $b_2 = b_1$ so that (a_2, b_2) is a maximal open interval that ends at $b_2 = b_1$ and that is a subset of F_2^c. This implies that $a_2 \in F_2$. Consequently $a_2 \neq a_1$. If $a_2 > a_1$ then denote $O_2 = (a_2, b_2)$. If $a_2 < a_1$, then ignore F_2, and repeat the same procedure with F_3.

Let $i_2 \geq 2$ be the smallest integer such that F_{i_2} satisfies the following: the maximal open interval (a_2, b_2) that ends at $b_2 = b_1$, and that is a subset of $F_{i_2}^c$, is such that $a_2 > a_1$. This must happen or some i_1 since the family of sets F_i covers I. In this case we have that $a_2 \in F_{i_2}$ (and so $a_2 \notin F_i$ for all $i \neq i_2$). With this new a_2, we denote $O_2 = (a_2, b_2)$. Note again that $a_1 < a_2 < b_2 = b_1$.

Now focus on a_2. It is in the open sets F_i^c for all $i \neq i_2$. Find the smallest $i_3 > i_2$ such that the set F_{i_3} satisfies the following: the maximal open interval (a_3, b_3) that starts at $a_3 = a_2$, and that is a subset of $F_{i_3}^c$, is such that $b_3 < b_2$. By an argument symmetric to the one used in the previous two paragraphs, this i_3 exists. Denote $O_3 = (a_3, b_3)$. Comparing O_1 with O_3 we see that $a_1 < a_3 < b_3 < b_1$.

Iterating this procedure we get a countable list of intervals $O_{2j+1} = (a_{2j+1}, b_{2j+1})$ such that $a_{2j+1} < a_{2j+3} < b_{2j+3} < b_{2j+1}$. By Exercise 15 (b), Section 2.2, we have $\bigcap_{i=1}^{\infty} O_{2i} \neq \varnothing$. It follows from our construction that $\bigcap_{j=1}^{\infty} F_i^c \supset \bigcap_{j=0}^{\infty} O_{2j+1}$ and so $\bigcap_{j=1}^{\infty} F_i^c \neq \varnothing$. By de Morgan laws, this gives $\bigcup_{i=1}^{\infty} F_i \neq I$, a clear contradiction.

6.2 Properties of Connected Spaces

Solutions to the odd-numbered exercises.

---

1. Show that X is not homeomorphic to Y.

(a) X is the subspace of $\mathbb{R}$ made of all irrational numbers, Y is $\mathbb{R}$.

(b) X is $\mathbb{R}$, Y is $\mathbb{R}^2$.

(c) X is the letter X, Y is the letter Y (both considered as subspaces of $\mathbb{R}^2$).

(d)

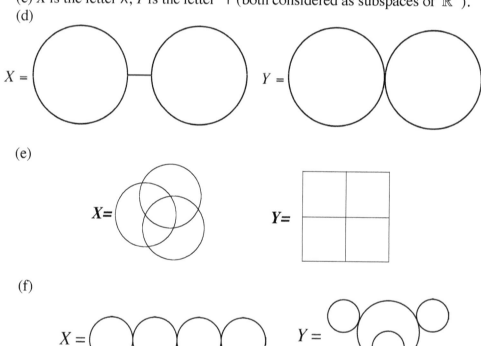

(e)

(f)

Illustration 6.13

Solution:

(a) X is disconnected, Y is connected; and see Proposition 1.

(b) Removing a point in $\mathbb{R}$ leaves a disconnected space, while that is not the case after removing a point in $\mathbb{R}^2$. The details of the argument are similar to what has been done in Example 1 in this section.

(c) Removing the crossing point in X leaves 4 components. No such point in Y.

(d) There are many points A in X such that $X \setminus \{A\}$ is disconnected. These points are sent by any homeomorphism to distinct points B of Y such that $Y \setminus \{B\}$ is disconnected. However, there is only one such point B in Y.

(e) We can remove 6 points of X without disconnecting (see Figure 9); only 4 in Y.

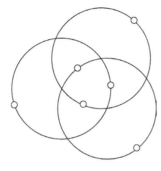

Figure 9

(f) Take three points from the big circle in Y in the arcs between the touching points (one from each such arc). Get three components, each one of them is a circle with a whisker. No combination of three points in X, after taking them our, leaves 3 components in X homeomorphic to circles with whiskers.

3. Suppose A is component of X and suppose $f : X \to Y$ is continuous. Show through a counterexample that $f(A)$ need not be a component of the subspace $f(X)$ of Y.

Solution. Take X to be the union space $[0,1] \cup [2.3]$, $Y = \mathbb{R}$ and define $f : X \to Y$ as

follows: $f(x) = \begin{cases} x & \text{if } x \in [0,1] \\ x-1 & \text{if } x \in [2,3] \end{cases}$. Then f is continuous but the image of the

component $[0,1]$ of X is not a component of $f(X) = [0,2]$.

5. Let X and Y be spaces, and let $A \subset X$, $B \subset Y$. Show that A is a component in X and B is a component in Y if and only if $A \times B$ is a component in the product space $X \times Y$.

Solution. Suppose A is a component in X and B is a component in Y. That $A \times B$ is connected follows from Theorem 5. If it is not maximal and if C is a connected subset properly containing $A \times B$ then ether $p_1(C)$ properly contains A or $p_2(C)$ properly contains B. In both cases we get a contradiction since both $p_1(C)$ and $p_2(C)$ are connected (as images of connected sets).
    Conversely, suppose $A \times B$ is a component in $X \times Y$. Suppose $A \subset U$ and U is connected. Then $U \times B$ is connected (since U and $B = p_2(A \times B)$ are connected. But $A \times B \subset U \times B$ and the maximality of $A \times B$ imply that $A \times B = U \times B$ and so $A = U$.

7. Use properties of connectedness to show that there is no continuous mapping $\mathbb{R} \to \mathbb{R}$ that sends the rational numbers to irrational numbers and irrational numbers to rational numbers.

Solution. Suppose such a mapping exists. Take any interval $[a,b]$. Since f is continuous, it preserves connectedness, and since intervals are the only connected subsets of $\mathbb{R}$, $f([a,b])$ is also an interval. There are uncountably many irrational numbers in the interval $f([a,b])$, and they are all images of rational numbers in $[a,b]$. So, there are uncountably many rational numbers in $[a,b]$. Contradiction.

Solution 2. Observe that $f(\mathbb{R})$ is countable. So, by exercise 13 in 6.1 it is disconnected. But f is continuous and $\mathbb{R}$ is connected, so $f(\mathbb{R})$ should be connected.

9. Show that if $f:\mathbb{R}^2 \to \mathbb{R}$ is continuous and if both 1 and –1 are images under f of some points in $\mathbb{R}^2$, then $f^{-1}(0)$ is uncountable.

Solution. Take the zero out of the interval $\mathbb{R}$. Then we get a disconnected space. So, since both –1 and 1 are hit by something, the inverse image of that space under f is also disconnected. So, $f^{-1}(0)$ is a curve that disconnects the plane and so, by the previous exercise (Exercise 8), it is uncountable.

11. An involution $f:X \to X$ is a continuous map such that $f \circ f = identity$. (So, involutions are homeomorphisms that are self-inverse.) Show that if $f:\mathbb{R} \to \mathbb{R}$ is an involution, then f must have a fixed point.

Solution. Choose $x \in \mathbb{R}$ and consider $f(x)$. If $f(x) = x$ then we are done. Otherwise, we may assume that $x < f(x)$ (the other case is symmetric). Then the interval $(x, f(x))$ is a component of $\mathbb{R} \setminus \{x, f(x)\}$, so f sends it either onto a component of $\mathbb{R} \setminus [x, f(x)]$ or onto itself. Assume that $f((x, f(x))) = (f(x), \infty)$. Then the third component, $(-\infty, x)$, is sent onto itself. However this violates the continuity of f since point close (and to the left of x) must go to points close to $f(x)$. The case when $f((x, f(x))) = (-\infty, x)$ is similar. The only remaining possibility is that $f((x, f(x))) = (x, f(x))$. Then $f([x, f(x)]) = [x, f(x)]$. Since $[x, f(x)]$ is homeomorphic to $[0,1]$, and it follows from the previous exercise (Exercise 10) that f much have a fixed point in $[x, f(x)]$.

13.   (a) Show that if $f:S^1 \to S^1$ is antipode-preserving, then f is onto. (A mapping $f:S^1 \to S^1$ is **antipode preserving** if $f(x) = -f(-x)$ for every $x \in S^1$.)
        (b) Show that if $f:S^1 \to S^1$ is an embedding than it is a homeomorphism.

Solution. (a) (Recall that we are assuming that all mappings are continuous, unless otherwise stated.) Suppose $x \in S^1$ is not in the range of f. Then, since f is antipode-preserving, neither is $-x$. These two points split S^1 into two components, A and B. The assumption that f is antipode-preserving now guarantees that $f(S^1) \cap A \neq \emptyset \neq f(S^1) \cap B$. So, $f(S^1)$ has at least two components, contradiction Proposition 1.

(b) We need to show that f is onto. Suppose it is not. Suppose there is no x such that $f(x) = y$, for some $y \in S^1$. Then the range of f is within $S^1 \setminus \{y\}$. But an easy argument regarding connectedness guarantees that no subset of (essentially) the interval $S^1 \setminus \{y\}$ is homeomorphic to S^1. On the other hand S^1 is homeomorphic to $f(S^1)$, and we have a contradiction.

6.3 Path Connected Spaces

---

1. Suppose $X \subset Y \subset Z$ and Z is a space. Show that X is path connected as a subset of Z if and only if it is path connected as a subset of the subspace Y of Z.

Solution. $\Rightarrow$ This follows from the following basic exercise (that we leave to the reader): Let Y be a subspace of Z, let W be any space, and let $f : W \to Y$ be any mapping. Then f is continuous if and only if $in \circ f : W \to Z$ is continuous, where $in : Y \to Z$ is the inclusion.

3. In Illustration 6.22 we show the subspace X of $\mathbb{R}^2$ consisting of the points on the spiral $r = 1 + \dfrac{10}{\varphi}$ in polar coordinates, $\varphi \in [2\pi, \infty)$, together with the points on the circle $r = 1$. Show that X is connected but not path connected.

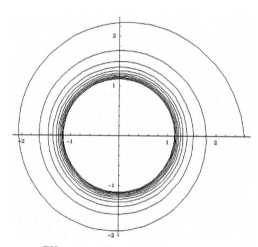

Illustration 6.22

Solution. The connectedness follows from Theorem 5, Section 6.1. Now, suppose there is a path α from a point $r = 1 + \dfrac{10}{\varphi_0}$ on the spiral, to the point $(1,0)$ on the inner circle. If a point $r = 1 + \dfrac{10}{\varphi_1}$, $\varphi_1 > \varphi_0$, is not in $\alpha([0,1])$, then it is easy to see that $\alpha([0,1])$ is disconnected, contradicting Proposition 1, in section 6.2. Hence each point $r = 1 + \dfrac{10}{\varphi_1}$, $\varphi_1 > \varphi_0$, is in $\alpha([0,1])$. Denote $t_0 = \sup\{t \in [0,1] : \alpha(s)$ is on the curve $r = 1 + \dfrac{10}{\varphi}$ for all $s < t\}$.

Suppose $\alpha(t_0)$ is on the spiral $r = 1 + \dfrac{10}{\varphi}$, $\varphi \geq \varphi_0$. Choose $t_1 > t_0$ such that $\alpha([t_0, t_1])$ is completely within the spiral. It follows from the definition of t_1 that there is $s \in (t_0, t_1)$ such that $\alpha(s)$ is not on spiral. This contradicts the connectedness of $\alpha([t_0, t_1])$. Hence $\alpha(t_0)$ is on the circle $r = 1$. Since each point P on $r = 1$ is an accumulation point for the spiral, there is a sequence of points $\alpha(s_1), \alpha(s_2), \alpha(s_3), \ldots$ in the spiral and converging to P. In particular, this is true if P is distinct from $\alpha(t_0)$.

However, this is not possible, since the continuity of α forces any $\alpha(s_1), \alpha(s_2), \ldots$ to converge to $\alpha(t_0)$.

5.   (a) Let A_i, $i \in I$, be path connected subsets of a space X and suppose $\bigcap_{i \in I} A_i \neq \varnothing$.

Show that $\bigcup_{i \in I} A_i$ is a path connected subset of X.

(b) Let A_0 and A_i, $i \in I$, be path connected subsets of a space X and suppose

$A_0 \cap A_i \neq \varnothing$ for every $i \in I$. Then $A_0 \cup \left(\bigcup_{i \in I} A_i \right)$ is path connected.

(c) Let A_i, $i = 1,2,\ldots$, be path connected subsets of a space X such that

$A_i \cap A_{i+1} \neq \varnothing$ for every $i = 1,2,\ldots$. Show that $\bigcup_{i=1}^{\infty} A_i$ is path connected.

Solution. (a) Choose any $x,y \in \bigcup_{i \in I} A_i$. So, $x \in A_j$, $y \in A_k$, for some $j,k \in I$. Since, by

assumption, $A_j \cap A_k \neq \varnothing$, there is $z \in A_j \cap A_k$. Let α be a path in A_j from x to z, and let

β be a path in A_k from z to y. Then $\alpha\beta$ is a path in $\bigcup_{i \in I} A_i$ from x to y.

(b) Similar to part (a): to find a path between two points in $A_0 \cup \left(\bigcup_{i \in I} A_i \right)$, find

paths to and from a point in A_0, then multiply them.

(c) Essentially the same as the proof of Proposition 3.

7. An *ambient path* in a space X starting from $P \in X$ and ending at $Q \in X$ is a mapping $h : X \times I \to X$ such that the restrictions $h\big|_{X \times \{t\}}$ is a homeomorphism for every t in I, such that $h\big|_{X \times \{0\}} : (x,0) \mapsto x$ for every $x \in X$, and such that $h\big|_{X \times \{1\}}$ sends P to Q. A space is *ambient path connected* if for every two points P and Q in X there is an ambient path in X from P to Q.

(a) Show that every ambient path connected space is path connected.
(b) Show that the (path connected) subspace of $\mathbb{R}^2$ depicted in Illustration 6.23 is not ambient path connected.

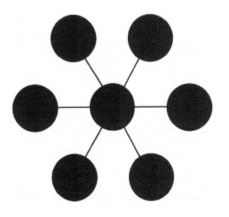

Illustration 6.23.

Solution. (a) Suppose X is ambient path connected. Choose any two points $P, Q \in X$. Then there is an ambient path $h : X \times I \to X$ from P to Q. Denote by f the homeomorphism $[0,1] \to \{P\} \times [0,1]$ defined by $f(t) = (P, t)$. Then $\left(h \big|_{\{P\} \times I} \right) \circ f$ is a path in X from P to Q.

    (b) Denote the object in the illustration by X (considered as a subspace of $\mathbb{R}^2$). Observe that every homeomorphism of X sends one of the seven disks into one of these disks, and that there is no homeomorphism $X \to X$ sending (say) the rightmost disk D_r onto the center disk D. To justify the latter claim note that $X \setminus D_r$ is (path) connected, and that $X \setminus D$ has six (path) components. Now choose a point $P \in D_r$ and choose a point $Q \in D$. Then there is no ambient path from P to Q, since if h is such, then $h \big|_{X \times \{1\}}$ must send D_r onto D.

6.4 Path Connected Spaces: More Properties and Related Matters

---

Solutions of the odd-numbered exercises.

---

1. Describe the path components in $\mathbb{R}$ with the half open interval topology.

Solution. Points.

3. Show that path components of open subsets of $\mathbb{R}^2$ are open.

Solution. Let A be an open subset of $\mathbb{R}^2$, and let C be a path component of A. Suppose C is not open. Then there is a point $x \in C$ that is not an interior point for C. Since $x \in A$ and since A is open, there is an open disk D such that $x \in D \subset A$. Since both C and D are path connected, and since $x \in D \cap C$, it follows from, say, Proposition 3, in 6.3, that $C \cup D$ is path connected, $C \subsetneq C \cup D \subset A$ contradicting the assumption that C is a path component of A.

Note 1: Virtually the same argument works for $\mathbb{R}^n$ in place of $\mathbb{R}^2$.
Note 2: This is a special case of the Proposition 1 in Section 6.5.

5. Consider the set $\mathbb{R}^2$ equipped with the post office metric (Section 2.1) with the designated point (the post office) being the origin. Denote by X the metric space generated by that metric.
    (a) Which of the following mappings from $[0,1]$ into X is a path in X from $(1,0)$ to $(0,1)$?

$$\alpha(t) = \begin{cases} (1-2t,0) & \text{if} \quad 0 \le t \le \frac{1}{2} \\ (0,2t-1) & \text{if} \quad \frac{1}{2} \le t \le 1 \end{cases}$$

$$\beta(t) = (1,0) + t(-1,1)$$

$$\gamma(t) = \begin{cases} (1,0) & \text{if} \quad t=0 \\ (0,0) & \text{if} \quad 0 < t < 1 \\ (0,1) & \text{if} \quad t=1 \end{cases}$$

    (b) Describe the path components of X.

Solution. (a) α is not continuous: the open ball B around the point $(1,0)$ and of radius (say) $\frac{1}{2}$ with respect to the post office metric, contains only the point $(1,0)$, and $\alpha^{-1}(B) = 0$ is not open in the interval $[0,1]$. The same argument applies to β and γ, hence β and γ are also discontinuous.

(b) Suppose a subset of X has two points. Then one of them is not the origin. Denote it by x. We already noted that if d is the Euclidean distance from x to the origin, then the open ball $B(x,r)$, where r is any positive number $< d$, and where the distance comes from the post office metric, contains only the point x. Consequently the two open sets $B\left(x,\frac{d}{2}\right)$ and $\overline{B\left(x,\frac{d}{4}\right)}^c$ disconnect X. We conclude that singletons are the only (path) connected subsets of X, and thus they are the (path) components of X.

7. Show that every space with co-finite topology and over an infinite countable set is connected but not path connected.

Solution. It is obvious that such a space is connected (every two proper open sets are not disjoint). If f is a path in such a space, then the inverse images of singletons are closed (since singletons are closed in co-finite topology and since f is continuous). So, the interval $[0,1]$ would be a disjoint union of countably many (more than 1) closed sets, which is not possible (Exercise 16 in Section 6.1).

9. Show that there are no two disjoint subsets A and B of $\mathbb{R}$, and a bijection $f : A \to B$, such that the space $\mathbb{R}_f$ (obtained by identifying A and B along f) is homeomorphic to $\mathbb{R}$.

Solution. Choose $a \in A$ and denote $f(a) = b$ ($b \in B$). We may suppose that $a < b$. If there is $x \notin A \cup B$ such that $a < x < b$, then taking it out of $\mathbb{R}_f$ will leave it path connected. This is obvious in case $A = \{a\}$ and $B = \{b\}$: there is a path α between any two points in $\mathbb{R}_{a=b}$. For larger A and B it suffices to compose α with the projection $\mathbb{R}_{a=b} \to \mathbb{R}_f$ to get a path between any two points in $\mathbb{R}_f \setminus \{x\}$. Since $\mathbb{R}$ minus a point is disconnected, the claim of the exercise follows for this case.

So, we may suppose that the interval $[a,b]$ is a subset of $C = A \cup B$. Bearing in mind the symmetry between the sets A and B (and mappings f and f^{-1}) – so that we can interchange their roles in the cases that follow – we have two cases.

Case 1: There is $a' \in [a,b] \cap A$ such that $b' = f(a')$ is not in $[a,b]$. We may suppose that $b' > b$ (the other case is symmetric, or at least similar). Choose $c \in [a,b]$ (see the figures below where we depict the space $\mathbb{R}_{a=b}$). Since $[a',b] \subset [a,b] \subset C$, we have that c is in C. By symmetry, we may suppose that $c \in A$. Consider the point $d = f(c)$. If $b < d < b'$ (Figure 10 top left) then take out c and d from $\mathbb{R}_{\substack{a=b \\ a'=b'}}$ and we are left with a path connected space; the preceding repetitive argument applies and so $\mathbb{R}_f$ minus $[c]=[d]$ is path connected (so, it could not be homeomorphic to $\mathbb{R}$). If $d > b'$ (Figure 10, top-right) then take out a' and b' from $\mathbb{R}_{\substack{a=b \\ c=d}}$ and apply the same argument. And finally, if $d < a$

(Figure 10, bottom) then take out a and b from $\mathbb{R}_{\substack{c=d \\ a'=b'}}$ and use again the same argument.

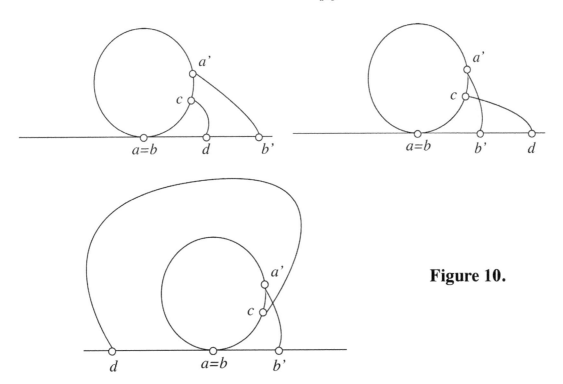

Figure 10.

Case 2: Now, suppose that for every $a' \in [a,b]$, b' is also in $[a,b]$. Consider a pair a_1, $b_1 = f(a_1)$, and choose $a_2 \in [a_1, b_1]$ (or $a_2 \in [b_1, a_1]$ if $b_1 < a_1$; we will neglect this symmetric case since it does not affect our argument). If $b_2 = f(a_1)$ (or $b_2 = f^{-1}(a_1)$; we will not pay attention to this either) is out of $[a_1, b_1]$, then taking out any one of $\{a_1, b_1\}$ and $\{a_2, b_2\}$ will leave the rest path connected, so that, as before, $\mathbb{R}_f$ could not be $\mathbb{R}$. So, if $a_2 \in [a_1, b_1]$, then $b_2 \in [a_1, b_1]$ too. Now, choose carefully: choose a_2 to be in the middle of $[a_1, b_1]$, so that the length of $[a_2, b_2]$ is half or less the length of $[a_1, b_1]$. Then choose a_3 in the middle of $[a_2, b_2]$ so that $[a_3, b_3]$ is at least twice shorter than $[a_2, b_2]$, etc. Note again that, say, since a_3 is chosen in $[a_2, b_2]$, b_3 must also be there, or else we are reverting to the case covered at the beginning of this paragraph. We repeat that, despite the notation, we are not assuming that $a_2 < b_2$, $a_3 < b_3$, etc. So, the end points of the intervals we are introducing may need to be reversed; this does not affect anything. Now: since $\lim_{n \to \infty} |a_n - b_n| = 0$, and since we have a nested sequence of intervals, Cantor theorem applies and we have a point x in the intersection of the intervals $[a_n, b_n]$. This x is then not paired with any other number, and so we are in Case 1. Done.

11. Find a space X that is
   (i)   Path connected but not disk-connected.
   (ii) Circle-connected but not sphere-connected
   (iii) Sphere-connected but not torus-connected.

Solution. (i) two disks with a line segment joining them;
         (ii) a sphere with a thick wall;
         (iii) a solid torus.

13. Prove Proposition 4.

Solution. Using the notation from the definitions of (isotopically) S-connected spaces, we merely observe that if $F\big|_{X \times \{t\}}$ is a homeomorphism for every $t \in [0,1]$, then $F\big|_{S \times \{t\}}$ is an embedding for every $t \in [0,1]$, and we are done.

15. Prove that the space Y depicted in Illustration 6.33 is S^1-connected, but not isotopically S^1-connected.

Solution. That it is S^1-connected one can verifying by seeing, visualizing and using symmetry of arguments. It is not isotopically S^1-connected because there is no isotopy from a copy of S^1 in one of the disks to a copy of S^1 around the cylinder. Otherwise, there would be a homomorphism of Y sending the first copy to the second copy. But taking out the first copy will not disconnect Y, while that is not the case for the second copy.

6.5 Locally Connected and Locally Path Connected Spaces

---

1. Show that each of the following spaces is both locally connected and locally path connected.
  (i)   Indiscrete space
  (ii) Discrete space
  (iii) Half-disk topology.

Solution. By the observation preceding Example 3 it suffices to prove that all these spaces are locally path connected.

(i) The only open neighborhood of a point in an indiscrete space X is X itself. X is surely path connected because any mapping $[0,1] \to X$ is continuous.

(ii) Let U be a non-empty open subset of a discrete space X, and let x be a point in U. Then $\{x\}$ is a path connected open set such that $x \in \{x\} \subset U$.

(iii) Denote by X the half-disk topology. Let $x \in U \subset X$, and U open. Then there is a set B from the basis used to define the half-disk topology, such that $x \in B \subset U$. Such a set B is either an open disk, or an open half disk together with the center point. In both cases it is clear that B is path connected.

3. Prove that if U is an open subset of $\mathbb{R}^2$, then the components and path components of U coincide.

Solution. If V is a path component of U, than it is connected, hence the corresponding component W contains V as a subset. Now, U is certainly locally path connected; for every point, just take the disk around that point that is within U. Hence (according to Proposition 2) each component of U is open. Hence V is open. Using he same argument we used for U we conclude that V must be locally path connected. So, V is connected and locally path connected; hence it is path connected (Proposition 3), and we have that V is a subset of W. So $W=V$.

5.   (a) Show that local (path) connectedness is not hereditary.
    (b) Show that if X is a locally (path) connected space, and A is an open subspace of X, then A is also locally (path) connected.

Solution. (a) is obvious – take horizontal lines $y = \frac{1}{n}$ together with the x-axis, and consider it as a subspace of the plane.

(b) Choose $a \in A$, and let U be an open subset of A containing a. We are searching for a (path) connected open subset V of A such that $a \in V \subset U$. Since A is open in X, it easily follows that U is also open in X. Since X is locally (path) connected, there is an open (path) connected subset V of X such that $a \in V \subset U$. But then, since U is a subset of A, so is V. Moreover, it is clear that V is also open in A (since A is open in X), and we are done.

7. Show that if I is finite and each of the spaces X_i are locally connected (locally path connected), then the space $\prod_{i \in I} X_i$ is locally connected (locally path connected, respectively).

Solution. Take a point and an open neighbourhood U of it in $\prod_{i \in I} X_i$. Project it and take open neighbourhoods U_i of the projection points. Then $\prod_{i \in I} U_i$ is an open connected neighbourhood of the starting point. Similar argument for local path connectedness.

9. Let $X/_\sim$ be a quotient space of a locally connected space X, and let $q : X \to X/_\sim$ be the quotient map.

    (a) Show that if C is a component of $X/_\sim$ then $q^{-1}(C)$ is a union of components of X. [Hint: show that if for every $x \in q^{-1}(C)$ the component of x in X is a subset of $q^{-1}(C)$].

    (b) Show that $X/_\sim$ is locally connected. [Hint: use Proposition 1.]

Solution. **(a)** Take $x \in q^{-1}(C)$ and let K be the component of x in X. Since $q(K)$ is connected and since $q(x)$ is in both C and $q(K)$, it follows that $q(K) \subset C$. So $q^{-1}(q(K)) \subset q^{-1}(C)$, and since $K \subset q^{-1}(q(K))$, we have that $K \subset q^{-1}(C)$.

    (b) Let U be an open set in $X/_\sim$ and let C be a component of U. By (a) $q^{-1}(C)$ is a union of components of X. By Proposition 1 each of these components is open. So, $q^{-1}(C)$ is open in X, so C is open in $X/_\sim$. Proposition 2 now implies that $X/_\sim$ is locally connected.

Chapter 7: Compactness and Related Matters

7.1 Compact spaces: definition

---

1. Which are compact subsets of a discrete space X? What if X is indiscrete?

Solution. A discrete space X is compact if and only if $|X|$ is finite. An indiscrete space X is always compact.

3. Prove Proposition 2: Suppose X is a subspace of Z. A subspace Y of the space X is compact if and only if it is compact as a subspace of Z.

Solution. Let Y be a compact subspace of X. Let $\mathcal{U}$ be an open cover of Y in Z. The intersections with X make an open cover in X. Extract a finite subcover and then take the corresponding open sets in Z. Conversely, suppose Y is compact in Z. Take an open cover in X. The open sets in this cover are intersections of open sets in Z and X. These open sets in Z make a cover for Y in Z. Extract finite subcover, then intersect each open set of that subcover with X.

5. Show that the Sorgenfrey line is not compact.

Solution: Recall that the Sorgenfrey line is the space over $\mathbb{R}$ generated by intervals of type $[a,b)$. We use Proposition 5: $(0,\frac{1}{n}]$, $n=1,2,\dots$, is a family of closed sets with the finite intersection property, yet the intersection of all these is empty.

7.   (a) Show that the interval $[0,\infty)$ is not compact.
         (b) Show that $[0,\infty)$ is not homeomorphic to the circle. (Hint: you may view the circle as being obtained by gluing the end points of the interval $[0,1]$.)

Solution.
         (a) $\{(0,n):n=1,2,\dots\}$ is an open cover of $[0,\infty)$ with no finite subcover.

(b) Every circle is compact, being a quotient space of a compact space. The claim now follows from Proposition 4.

9. Prove that if X is a compact space, then the diagonal $\{(x,x); x \in X\}$ is a compact subset of $X \times X$.

Solution. The mapping $x \to (x,x)$ is continuous mapping from X onto the diagonal. The claim now follows from Proposition 4.

11. Show that if a set X is well ordered, and if there exist the largest element m in X then X with the order topology is a compact space.

Solution. Let $\mathcal{U}$ be an open cover for X made of intervals. Then $m \in (a_1, m]$ for some $(a_1, m] \in \mathcal{U}$. Take this a_1 to be the smallest of all a such that $(a, m] \in \mathcal{U}$ (using well ordering). If $(a_1, m] = X$, we are done. Otherwise, $a_1 \in (a_2, b_2)$ for some $(a_2, b_2) \in \mathcal{U}$. Again take a_2 to be the smallest such. Note that $a_2 < a_1$, and, in particular, they are distinct. Keep going. Get a sequence $\{a_n\}$, $a_1 > a_2 > \cdots$. There is no infinitely decreasing sequence in a well ordered set (since the smallest element could not belong to that sequence), So, at some point we arrive at a_m such that there are no elements in X that are smaller than a_m. But then $(a_1, m], (a_2, b_2), ..., (a_m, b_m)$ covers all X and we are done.

13. Show that every compact metric space is complete.

Solution. Suppose X is compact. Hence, by Exercise 12, every sequence has a convergent subsequence. Let (x_i) be a Cauchy sequence in X; we will prove that (x_i) must converge. Let (x_{i_j}) be a convergent subsequence of (x_i), and suppose it converges to some number a. Hence for every $\epsilon > 0$, there is $N \in \mathbb{Z}^+$ such that if $i_j > N$, then $\left| x_{i_j} - a \right| < \frac{\epsilon}{2}$. Since (x_i) is Cauchy, for the same $\frac{\epsilon}{2} > 0$, there is $M \in \mathbb{Z}^+$, such that if $n, m > M$, then $\left| x_n - x_m \right| < \frac{\epsilon}{2}$. Choose $K = \max\{N, M\}$. Choose any $i_j > K$. Then for every $k > K$ we have the following:

$$\left| x_k - a \right| = \left| x_k - x_{i_j} + x_{i_j} - a \right| \leq \left| x_k - x_{i_j} \right| + \left| x_{i_j} - a \right| < \frac{\epsilon}{2} + \frac{\epsilon}{2} = \epsilon, \text{ establishing that } (x_i) \text{ converges}$$

to a as well.

15. (a) Show that there is no homeomorphism $f : \mathbb{R}^2 \to \mathbb{R}^2$ such that $f(A) = B$ (see Illustration 7.3).

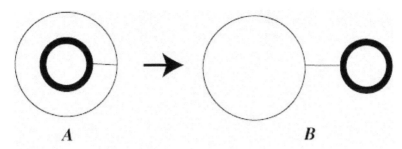

Illustration 7.3. The subspace A of $\mathbb{R}^2$ is shown to the left (all black points); B is to the right.

(b) Describe (visually or otherwise) a homeomorphism $g : \mathbb{R}^2 \to \mathbb{R}^2$ such that $g(A) = B$ (see Illustration 7.4).

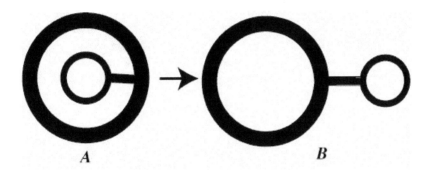

Illustration 7.4.

Solution. (a) Such a homeomorphism must send the complement of A to the complement of B. Because of connectedness the horizontal line segment of A must go to the horizontal line segment of B. Because of that and because of continuity and connectedness the region R of A between the circles must go to the region S of B outside the circles. So, since $f(A) = B$ (and because continuous maps send boundary to boundary) it follows that the closure $\overline{R}$ of R must go to the closure $\overline{S}$ of S. But then $\overline{R}$ is compact, while $\overline{S}$ is not. Contradiction.

(b) Starting from A, a slow animation pushing the smaller circle to the right, inside the thick boundary of the larger circle, then out of it to the right, ending at B, gives a bunch of homeomorphism, the composition of which establishes the desired homeomorphism. This is illustrated in the following five frames (Figure 11), where each arrow indicates an evident homeomorphism.

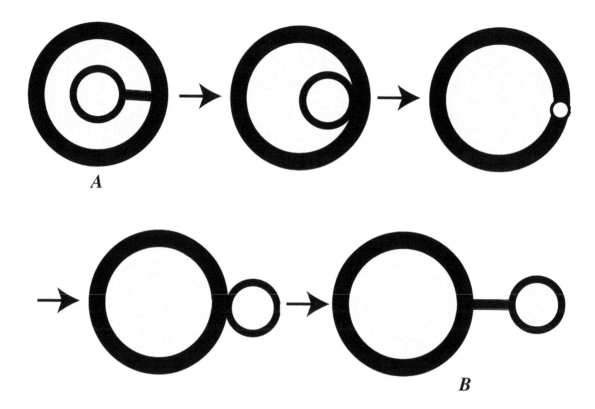

Figure 11.

17. Show that if $\{A_j : j \in 1,2,...\}$ is a family of compact non-empty subsets of a

Hausdorff space X such that $A_1 \supset A_2 \supset ... \supset A_j \supset ...$, then $\bigcap_{j=1}^{\infty} A_j \neq \varnothing$. Show that if

'Hausdorff" is omitted than the new statement can fail.

Solution. Every compact subspace of a Hausdorff space must be closed. This is left as an exercise at this stage; the proof is supplied in Proposition 2, Section 7.2. Hence each space A_j is closed. Since we have assumed that A_j-s are not empty, it follows easily that the family $\{A_j : j \in 1,2,...\}$ has the finite intersection property. It then follows from the observation that each A_j, $j \geq 2$, is a closed subset of the compact (and Hausdorff) space

A_1, and from Proposition 5, that $\bigcap_{j=1}^{\infty} A_j \neq \varnothing$.

　　　Second part: Consider the co-finite topology over $\mathbb{N}$. First we prove that every open set in this topology is compact! Open sets are complements of finite sets. So, take any such open set $V = \mathbb{N} \setminus \{a_1,...,a_m\}$; choose an open cover of that set. Take any open set in that open cover. It avoids finitely many elements of $\mathbb{N}$, hence finitely many elements of V. Now choose one open set from the cover for each of these missing elements of V; that is a finite subcover. Notice that this topology is not Hausdorff.

Now, the family $\{\{n, n+1, ...\} : n \in \mathbb{N}\}$ is a nested family of compact sets, and their intersection is empty.

19. Show that every finite CW-complex is compact.

Solution. If X is the same as its 0-skeleton, then X is finite, hence it is compact. Suppose its k-skeleton is compact. Then the sum (disjoint union) of this k-skeleton and the $(k+1)$-cells that we attach (each of them compact subspace of an Euclidean space) is also compact. Since a quotient space of a compact space is compact, it follows that the $(k+1)$-skeleton is also compact. The result now follows by induction.

7.2 Properties of Compact Spaces

Solutions of the odd-numbered exercises.

---

1. Show that a finite union of compact subsets of a space X is a compact subset of X.

Solution. Let $A_1, A_2, ..., A_n$ be compact subsets of X and suppose $\mathcal{U}$ is an open cover of $\bigcup_{i=1}^{n} A_i$. Then $\mathcal{U}$ is an open cover of each A_i. Since these sets are compact, there is a finite subcover $\mathcal{U}_i$ of $\mathcal{U}$ covering A_i, $i = 1, 2, ..., n$. Then $\bigcup_{i=1}^{n} \mathcal{U}_i$ is a finite subcover of $\mathcal{U}$ covering $\bigcup_{i=1}^{n} A_i$. Hence $\bigcup_{i=1}^{n} A_i$ is compact.

3. Show that compact subsets of a compact space need not be closed.

Solution. Take X to be indiscrete and A to be any proper non-empty subset of X. Then both A and X are surely compact, but A is not closed in X.

5. Show that the closure of a compact subspace of a space X need not be compact. (Hint: consider the Particular-point topology over an infinite set X, with open sets being all sets containing a fixed point of X.)

Solution. The closure of the set $\{p\}$, where p is the chosen point in the Particular-point topology, is all of X. The set $\{p\}$ is compact, but X is not, since the cover consisting of all two-element sets $\{p, x\}$, x ranges through X and is not p, has no finite subcover. If X *is* Hausdorff than compact subsets must be closed (Proposition 2), and so the closure of compact sets are surely compact.

7. Let X be a compact space, let Y be Hausdorff, and let $f : X \to Y$ be a bijection.
      (a) Show that if f is continuous that it is open.
      (b) Show that if f is open then it is continuous.

Solution. (a) Suffices to show that f is closed; it immediately follows from the assumption that f is a bijection that it is open. Take a closed subset of X. Then it is compact (Proposition 1). Hence its image is compact. But Y is Hausdorff, and so this image is closed (Proposition 2).

  (b) If f is open then f^{-1} is continuous and the rest follows from part (a).

9.   (a) Show that if A is a compact subset of a Hausdorff space X and if $x \in X$ is not in A, then there are two disjoint, open subsets U and V of X such that $A \subset U$ and $x \in V$.

  (b) Show that for every two disjoint compact subsets A and B of a Hausdorff space X, then there are two disjoint, open sets U and V such that $A \subset U$ and $B \subset V$.

  Solution. (a) Since X is Hausdorff, for every $a \in A$ there are open sets U_a and U_x^a around a and x respectively such that $U_a \cap U_x^a = \varnothing$. All such U_a-s, $a \in A$, form a cover of A. Extract a finite subcover $U_{a_1}, ..., U_{a_n}$. Then $U_{a_1} \cup U_{a_2} ... \cup U_{a_n}$ and $U_x^{a_1} \cap U_x^{a_2} \cap ... \cap U_x^{a_n}$ are as wanted.

  (b) By part (a) for every $b \in B$ there are disjoint open sets U^b and U_b such that $A \subset U^b$, and $b \in U_b$. Then $\{U_b : b \in B\}$ is an open cover of B. Extract a finite subcover $U_{b_1}, ..., U_{b_n}$. Then $U_{b_1} \cup U_{b_2} ... \cup U_{b_n}$ and $U^{b_1} \cap U^{b_2} \cap ... \cap U^{b_n}$ are as wanted.

11. Denote $X = \{(x,0) : 0 \le x \le 1\} \cup \bigcup_{s \in J} \{s\} \times [0,1]$, where J is a totally disconnected subset of the interval $[0,1]$ (see Illustration 7.18). Show that if $f : [0,1] \to X$ is a continuous mapping, then $f([0,1]) \cap \bigcup_{s \in J} \{s\} \times \{1\}$ is finite.

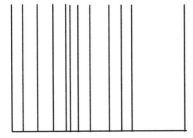

Illustration 7.18.

Solution. Take two distinct points $P = \{a\} \times \{1\}$ and $Q = \{b\} \times \{1\}$ (both top points of the space). Denote $A = \{a\} \times \{0\}$.

We first claim that every path from P to Q must hit A. Suppose otherwise, that there is a path f from P to Q avoiding A. Since the image of the unit interval is compact in the

Hausdorff space *X,* it must be closed. Hence, there is an open neighborhood around *A* disjoint from $f(I)$. Take a point $(c,0)$ in that neighborhood (in the bottom horizontal line), to the right of *A* if *Q* is to the right of *P,* to the left of *A* otherwise, and at distance to *A* less than the distance between *P* and *Q.* Since the vertical line segments intersect the bottom horizontal line at a totally disconnected subset of that horizontal line segment, there must be a point $(x,0)$ in the interval from $(a,0)$ to $(c,0)$ such that there is no vertical line segment in *X* above $(x,0)$. Then the vertical line through $(x,0)$ does not intersect $f(I)$ and the two half planes defined by that line would separate $f(I)$ into two non-empty open sets, which is not possible. We have a contradiction. Hence every path from *P* to *Q* must pass through *A.*

The rest is easy, for if $f(I)$ intersects infinitely many points at the top, then it must include all of the corresponding vertical line segments, and the inverse images under f^{-1} of the middle open sub-segments of these vertical line segments would make an infinite family of disjoint open subsets of *I.* That would violate the compactness of *I.*

7.3 Around Compactness; Lindelöf spaces

Solutions of the odd-numbered exercises.

---

1. Suppose $f : [a,b] \to \mathbb{R}$ is continuous. Show that $f([a,b])$ is a closed and bounded interval.

Solution. By Proposition 4 in 7.1, $f([a,b])$ is compact. By Heine-Borel Theorem (Theorem 6 in 7.2), $f([a,b])$ is closed and bounded. By Proposition 1 in 6.2, $f([a,b])$ is connected. By Proposition 2 in 6.1 it must be an interval.

3. (a) Prove that a space X is countably compact if and only if for every countable collection $\{F_j : j = 1,2,...\}$ of closed subsets of X that has the finite intersection property,

$$\bigcap_{j=1}^{\infty} F_j \neq \varnothing .$$

(b) Let Y be a countably compact subspace of a space X, and let $\{F_j : j = 1,2,3,...\}$ be a collection of closed nonempty subsets of Y such that

$F_1 \supset F_2 \supset F_3 \supset \cdots$. Show that $\bigcap_{j=1}^{\infty} F_j \neq \varnothing$.

Solution. (a) $\Rightarrow$ Suppose X is countably compact and let $\{F_j : j = 1,2,...\}$ be a family of closed subsets of X that has the finite intersection property. Suppose that $\bigcap_{j=1}^{\infty} F_j = \varnothing$. Then

$\left(\bigcap_{j=1}^{\infty} F_j \right)^c = X$, and so $\bigcup_{j=1}^{\infty} F_j^c = X$. This means that the family $\{F_j^c : j = 1,2,3,...\}$ is a countable open cover of X. Since we have assumed that X is countably compact, there is a finite subcover $\{F_{i_1}^c, F_{i_2}^c, ..., F_{i_n}^c\}$, and hence $\bigcup_{j=1}^{n} F_{i_j}^c = X$. After taking the complement on both sides this yields $\bigcap_{j=1}^{n} F_{i_j} = \varnothing$, contradicting the assumption that $\{F_j : j = 1,2,...\}$ has the finite intersection property. $\Leftarrow$ this requires an argument that traces the previous one in the opposite direction.

(b) is a consequence of (a).

5. Show that if $f : X \rightarrow Y$ is continuous, onto, and if X is Lindelöf, then so is Y.

Solution: Assume X is Lindelöf. Let $\{V_i : i \in I\}$ be an open cover of Y. Then $\{f^{-1}(V_i) : i \in I\}$ is an open cover of X. By assumption the last cover possess a countable subcover $\{f^{-1}(V_{i_j}) : j \in J\}$. Then $\{V_{i_j} : j \in J\}$ must cover all of Y, and we have found a countable subcover of $\{V_i : i \in I\}$.

7. Show that Lindelöf property is not hereditary.

Solution. Take any space that is not Lindelof, say, the discrete topology over the set of real numbers. Consider the one-point compactification: it is compact, hence Lindelof. But the original space is a subspace that, by assumption, is not Lindelof.

7.4 Bolzano, Weirstrass, Lebesgue and Jordan

Solutions of the odd-numbered exercises.

---

1. Show that every closed subspace of a BW-space is a BW-space.

Solution: Let X be a BW-space and let Y be a closed subspace of X. Let A be an infinite subset of Y. Then A is an infinite subset of X, and since X is a BW-space, there must be an accumulation point p for A such that $p \in X$. This p is then an accumulation point for Y, and since Y is closed, it must be that $p \in Y$. Hence Y is a BW-space too.

3. Let X be an uncountable space equipped with the co-countable topology, and let $Y = [0,1]$ be with the indiscrete topology. Show that $X \times Y$ is a BW space that is not compact.

Solution. Take an infinite set A in $X \times Y$. Suppose $(a,y) \in A$; then every open neighborhood of (a,y'), $y' \neq y$, must be of type $U \times [0,1]$, with $a \in U$, and so it contains (a,y). So (a,y') must be an accumulation point. Hence $X \times Y$ is a BW-space.

Take the any infinitely countable subset $B = \{b_1, b_2, \ldots\}$ of X, so that $U = X \setminus B$ is open in X. It follows that $U \times [0,1]$ is open in $X \times Y$. Look at the following open subsets of $X \times Y$: $U_1 = \big((X \setminus B) \cup \{b_1\}\big) \times [0,1]$, $U_2 = \big((X \setminus B) \cup \{b_1, b_2\}\big) \times [0,1], \ldots$. It is obvious that this collection of open sets covers $X \times Y$. It is also obvious that a finite cover cannot be extracted.

5. Let X be a Hausdorff BW-space, and let (F_i) be a sequence of closed, non-empty subsets of X such that $F_{i+1} \subset F_i$ for every $i \in \mathbb{Z}^+$. Show that $\bigcap_{i=1}^{\infty} F_i \neq \varnothing$.

Solution 1 (indirect). By Proposition 1, X is countably compact. The claim now follows from Exercise 3(b) in Section 7.3.

Solution 2 (direct). Choose a point $x_i \in F_i$. Get a sequence. If an element x appears infinitely many times in this sequence, then $x \in \bigcap_{i=1}^{\infty} F_i$ and so $\bigcap_{i=1}^{\infty} F_i \neq \varnothing$. Otherwise the sequence (x_i) has infinitely many distinct elements. Denote this set by A. Since X is a BW, A has an accumulation point y. Suppose $x \notin \bigcap_{i=1}^{\infty} F_i$. Then there is j such that $x \notin F_j$.

So x is in the open set F_j^c. The set $G = \{x_i : i = 1, 2, ...\} \cap (F_j^c \setminus \{x\})$ is a subset of $\{x_1, x_2, ..., x_{j-1}\}$, and so it is finite. Since X is Hausdorff, singletons are closed, and so G is also closed. Consequently G^c is open; thereby $U = G^c \cap F_j^c$ is also open. It contains x by construction, and it does not contain any element of A. Hence x is not an accumulation point for A. This is a contradiction. We conclude that $x \in \bigcap_{i=1}^{\infty} F_i$, and so $\bigcap_{i=1}^{\infty} F_i \neq \varnothing$.

7. A space X is **sequentially compact** if every sequence has a convergent subsequence.
    (a) Show that if X is sequentially compact then it is a BW-space
    (b) Show that if X is first countable then the converse is true.

Solution

(a) Suppose X is sequentially compact. Take any infinite subset A of X, and choose any infinite sequence $\{a_n\}$ of distinct elements of A. By assumption, it has a subsequence that converges to some $a \in X$. Then it is obvious that a must be an accumulation point for the set $\{a_n : n \in \mathbb{N}\}$, and so it is in A' too.

(b) Suppose now that X is a first countable, BW-space. Let (a_n) be any sequence of elements of X. If the sequence has only finitely many distinct elements, then it stabilizes, and so a subsequence made of the stabilizing element will converge. Suppose (a_n) has infinitely many elements. Take any subsequence (b_n) with distinct elements. Then, since X is BW, there is an accumulation point b for the set $\{b_n : n \in \mathbb{N}\}$. The rest now follows from Exercise 23 in 3.3.

9. Show that every compact subspace of the Sorgenfrey line is countable. [Hint: You may want to start with the identity map from the Sorgenfrey line onto the Euclidean line $\mathbb{R}$).]

Solution, Start with a compact subspace A of the Sorgenfrey line (S-line). Then the identity mapping from the S-line to the usual line, being continuous, sends A to a compact subspace of the usual line. This must be closed and bounded. So, all of A is in a closed interval $[a, b]$.

Suppose there is an infinite increasing sequence $\{c_n\}$ of elements in A converging in the usual topology to some $s \in \mathbb{R}$. Consider the family $\{[a, c_n) : n = 1, 2, ...\} \cup \{[s, b+1)\}$: it is an open (in the S-line) cover of A that possesses no finite subcover of A. Hence there is no such infinite sequence $\{c_n\}$. This implies that for every $a \in A$ there is an open interval $(a - \epsilon_a, a)$ containing no elements from A. Choose a rational number $q_a \in (a - \epsilon_a, a)$. Then $a \mapsto q_a$ defines a one-to-one mapping $A \to \mathbb{Q}$, which implies that the cardinality of A is at most the cardinality of $\mathbb{Q}$.

7.5 Compactification

Solutions of the odd-numbered exercises.

---

1. Show that the one-point compactification X_∞ of a space X is a topological space.

Solution. The only non-trivial part of the proof concerns unions of open sets of type (c) (as defined on page 175), and here we need the fact that closed subsets of compact sets are compact. Let $\{U_j : j \in J\}$ be a collection of open sets in X_∞ such that the point ∞ is in $A = \bigcup_{j \in J} U_j$. Hence at least one $U \in \{U_j : j \in J\}$ is of type (c). We wish to show that

$X \setminus A$ is closed and compact. We have: $X \setminus A = X \cap \left(\bigcap_{j \in J}(U_j)^c \right) = \bigcap_{j \in J} X \cap (U_j)^c$ which is

closed in X (being an intersection of closed sets). Observe that $X \setminus A \subset X \setminus U$, and since $X \setminus U$ is compact, it follows from Proposition 1 in 7.2 that $X \setminus A$ is compact as well, as desired.

3. Let X be the subspace of the unit circle S^1 consisting of the points in S^1 with rational polar angles (in degrees), and let $Y = X \setminus \{(0,1)\}$. Is X homeomorphic to the one-point compactification of Y?

Solution. No. The space X is not compact at all, since it is not closed. Not possible for subspaces of the Euclidean space. (Actually, there are infinitely many accumulation points in $\mathbb{R}^2 \setminus X$.

5. Find the one-point compactification of a discrete space X with countably many elements. Describe a subspace of $\mathbb{R}$ which is homeomorphic to the one-point compactification of $\mathbb{N}$.

Solution. We can take X to be $\mathbb{N}$ as a subspace of $\mathbb{R}$. Then X_∞ is a convergent sequence, together with the point of convergence playing the role of ∞. The simplest to see this is via the stereographic projection where ∞ will be the north pole, and the rest of $\mathbb{N}$ will form a sequence on the circle converging to the north pole. The space is homeomorphic to the subspace $S = \left\{ \dfrac{1}{n} : n \in \mathbb{N} \right\} \cup \{0\}$ of $\mathbb{R}$, with the obvious

homeomorphism f sending $n \in \mathbb{N}$ to $\dfrac{1}{n} \in S$, and sending ∞ to 0. For example, it is easy to see that open subsets of X_∞ of type (c) (as in the definition of X_∞) are finite subsets of $\mathbb{N}$ together with ∞, and that they correspond via f to open subsets of S.

7. Show that if X is a subspace of a compact space Y, then there is a compact subspace Z of Y such that $X \subset Z$ and such that X is dense in Z.

Solution. Take $Z = \overline{X}$, the closure taken in Y. This means that X is dense in Z. Now, Z is a closed subset of the compact Y, so it is compact.

9. Show that a locally compact space need not be compact.

Solution. Take any infinite set with the discrete topology.

11. (a) Show that a closed subset of a locally compact space is locally compact. [You might need Exercise 16, in 4.1.]
   (b) Show that an open subset of a locally compact space need not be locally compact. [Hint: $\mathbb{Q}$]

Solution.
   (a) Let F be a closed subset of a locally compact space X and let $x \in F$. We want to show that there is an open subset U of F and containing x such that $cl_F(U)$ is compact. What we do have is that there is an open subset V of X such that $cl_X(V)$ is compact. Take $U = V \cap F$. Then, by Exercise 16, in 4.1, $cl_F(U) = cl_X(V) \cap F$. Since F is closed in X, it follows that $cl_X(V) \cap F$ is closed in $cl_X(V)$. Now, $cl_X(V)$ is compact, and so (theorem) closed subsets of $cl_X(V)$ are also compact. So $cl_X(V) \cap F$ is compact (in $cl_X(V)$, hence everywhere), which means that $cl_F(U)$ is compact. And that is what we wanted.

   (b) Consider the one point compactification $\mathbb{Q}_\infty$ of the space $\mathbb{Q}$ ($\mathbb{Q}$ considered as a subspace of $\mathbb{R}$). Then $\mathbb{Q}_\infty$ is compact, hence locally compact. $\mathbb{Q}$ is open in $\mathbb{Q}_\infty$. However $\mathbb{Q}$ is not locally compact, since closed intervals in $\mathbb{Q}$ are not compact.

7.6 Infinite Products of Spaces and Tychonoff Theorem

Solutions of the odd-numbered Exercises.

---

1. Prove the converse of Tychonoff theorem: if $\prod_{i \in J} X_i$ is compact, then so every X_j, $j \in J$.

Solution. For every $j \in J$ the projection $\prod_{i \in J} X_i \to X_j$ is onto and continuous. The result now follows from Proposition 4 in 7.1.

3. Show that an infinite and countable discrete space X is never equal to the product space $\prod_{j \in J} Y_j$, where each Y_j is not homeomorphic to X.

Solution. X could not be the product space of (countably many) finite discrete spaces, since the former is not compact, while the latter is (as a product of compact spaces!). So, at least one of Y_j is infinite and countable. If it is not discrete, then it is easy to see that the product space would also not be discrete. If it is discrete, then it must be homeomorphic to X.

5. Consider the **Tychonoff cube** $X = \prod_{j \in I} I_j$, where each I_j is a copy of $I = [0,1]$. Let H be the subspace of X consisting of all non-decreasing functions $I \to I$ (H is the **Helly space**). Show that H is compact. [Hint: exhibit H as $f(X)$ for a continuous $f : X \to X$.]

Solution: Define $f : X \to X$ as follows: for every element $g : I \to I$ of X, the mapping $f(g) : I \to I$ is defined by $f(g)(x) = \sup\{g(t) : t \in [0,x]\}$. That this is onto H is obvious. That it is continuous follows from Exercise 19 in 2.2.

7. Let X_i, $i \in I$, be nonempty sets, let $y_i \notin X_i$ for every $i \in I$, and define topologies over each $Y_i = X_i \cup \{y_i\}$ by declaring open sets to be $\varnothing, X_i, \{y_i\}, Y_i$. Denote $Y = \prod_{i \in I} Y_i$, and let $p_i : Y \to Y_i . i \in I$, be the projections.

    (a) Use Tychonoff theorem to show that $C = \left\{ p_i^{-1}(y_i) : i \in I \right\}$ is not a cover of Y.

(b) Deduce Axiom of Choice; that is, show that there exists $f : I \to \bigcup_{i \in I} X_i$ such that $f(i) \in X_i$ for every $i \in I$.

Solution. (See for example http://at.yorku.ca/p/a/c/a/01.htm)

Start with a family X_i, $i \in I$ of (pairwise disjoint) nonempty sets. We want to ding a function $f : I \to \bigcup_{i \in I} X_i$, such that $f(i) \in X_i$ for every i.

Let $y_i \notin X_i$ for every i. Consider the spaces $Y_i = X_i \cup \{y_i\}$, equipped with the following topology $\tau = \{\emptyset, X_i, \{y_i\}, Y_i\}$. These are compact, and hence so is $Y = \prod_{i \in I} Y_i$. Look at the sets $C_j = \{y_j\} \times \prod_{\substack{i \neq j \\ i \in I}} Y_i$. They are all open. If they make a cover of Y, then, by Tychonoff, there will be a finite subcover $C_{j_1}, C_{j_2}, ..., C_{j_n}$. However, it is obvious that this is not a cover since, say, $x_{j_1}, x_{j_2}, ..., x_{j_n}$, (*anything* in *the remaining coordinates*), $x_{j_k} \in X_{j_k}$, is not in any of C_{j_k}-s. Hence the sets $C_j = \{y_j\} \times \prod_{\substack{i \neq j \\ i \in I}} Y_i$ do not make a cover of Y. So, there exists an element in f in Y that is out of each C_j. Each coordinate of this element thus must be in some X_j and we have achieved what we wanted.

Chapter 8: Separation Properties

8.1 The Hierarchy of Separation Properties

Solutions of the odd-numbered exercises.

---

1. Show that X is a T_0 space if and only if for every distinct $x, y \in X$, $\overline{\{x\}} \neq \overline{\{y\}}$.

Solution. $\Rightarrow$ Suppose X is a T_0 space. Then there is a neighbourhood around, say, x, that avoids y. It follows that x is not in the closure of y and so $\overline{\{x\}} \neq \overline{\{y\}}$.

$\Leftarrow$ Suppose for every $x \neq y$, we have $\overline{\{x\}} \neq \overline{\{y\}}$. So there is z in one of $\overline{\{x\}}, \overline{\{y\}}$ that is not in the other. We may assume that $z \in \overline{\{x\}}$ and $z \notin \overline{\{y\}}$. The latter implies that that there is a neighborhood U of z that avoids y. The former implies that this neighborhood must include x,

3. Let A and B be two disjoint subsets of a space X. Show that if there are open sets U and V that separate A and B, then $\{A, B\}$ is a separation of the subspace $A \cup B$ of X. Show that the converse is false in general.

Solution. $\Rightarrow$ Suppose the open sets U and V separate A and B. So $A \subset U$, $B \subset V$ and $U \cap V = \varnothing$. So, $A = A \cap U$, $B = B \cap V$, hence the sets A and B are open in $A \cup B$. Since $A \cap B = \varnothing$, it follows that $\{A, B\}$ is a separation of the subspace $A \cup B$ of X.

For the second claim, consider $X = \{a, b, c\}$ and declare open sets to be the empty set and every set containing c. Set $A = \{a\}$ and $B = \{b\}$. Then A and B are open in $A \cup B$ (as a subspace of X), and so $\{A, B\}$ is a separation of $A \cup B$. On the other hand A and B cannot be separated in X since every non-empty open set in X contains c.

5. Show that the Zariski topology over $\mathbb{R}^n$ is T_1 but not T_2.

Solution. Every two distinct points in $(a_1, a_2, ..., a_n), (b_1, b_2, ..., b_n) \in \mathbb{R}^n$ differ in at least one coordinate. Suppose $a_i \neq b_i$. Then the basis sets $U = \{(x_1, x_2, ..., x_n) \in \mathbb{R}^n : x_i = a_i\}^c$ and $V = \{(x_1, x_2, ..., x_n) \in \mathbb{R}^n : x_i = b_i\}^c$ are such that $(a_1, a_2, ..., a_n) \notin U$, $(b_1, b_2, ..., b_n) \in U$, and $(a_1, a_2, ..., a_n) \in V$, $(b_1, b_2, ..., b_n) \notin V$. This shows that the Zariski topology is T_1.

That the same topology is not T_2 follows from the fact that every two nonempty, distinct open set belonging to the defining basis of the Zariski topology have a nonempty intersection. Here is a proof: Let U be the complement of

$\{(x_1, x_2, ..., x_n) \in \mathbb{R}^n : p_1(x_1, x_2, ..., x_n) = 0\}$ and let V be the complement of

$\{(x_1, x_2, ..., x_n) \in \mathbb{R}^n : p_2(x_1, x_2, ..., x_n) = 0\}$. Suppose further that U and V are nonempty. So, the polynomials p_1 and p_2 are not trivial, meaning not all coefficients are 0. Now consider the polynomial $p = p_1 p_2$. It is also non trivial. Moreover, it is easy to observe that the complement of $\{(x_1, x_2, ..., x_n) \in \mathbb{R}^n : p(x_1, x_2, ..., x_n) = 0\}$ is contained in $U \cap V$. Finally, since p is not trivial, there is at least one point $(a_1, a_2, ..., a_n) \in \mathbb{R}^n$ such that $p(a_1, a_2, ..., a_n) \neq 0$.Consequently (prove the 'consequently') the complement of $\{(x_1, x_2, ..., x_n) \in \mathbb{R}^n : p(x_1, x_2, ..., x_n) = 0\}$ is nonempty, and so $U \cap V \neq \varnothing$.

7. Prove Theorem 3: Subspaces of T_i-spaces are T_i-spaces, $i = 1, 2, 3$.

Solution. We prove the claim for T_3 spaces; a similar (but easier) argument covers the other cases. So, let X be a T_3 space, and let Y be a subspace of X. Let F be a closed subset of Y, and choose any $y \in Y \setminus F$. By Proposition 2 in 4.1 there is a closed subset G of X such that $F = G \cap Y$. Since $y \notin F$, it follows that $y \notin G$. Since X is T_3, there are disjoint open sets U and V, such that $G \subset U$ and $y \in V$. Then the sets $U \cap Y$ and $V \cap Y$ are open in Y, disjoint, and $F \subset U \cap Y$, $y \in V \cap Y$. Hence Y is T_3.

9. (a) Find a normal space X and a quotient space $Y = X/\sim$ such that Y is not Hausdorff.

(b) Find a normal space X and a quotient space $Y = X/\sim$ such that Y is not a T_0-space.

Solution. **(a)** Chose X to be two disjoint copies of $\mathbb{R}$ (that is, the sum of $\mathbb{R}$ with itself); to get Y identify each point of one copy with the corresponding point in the other copy, except for one pair of such points.

 (b) (This, of course, solves (a) too). Take $\mathbb{R}$ and partition it into two sets: one containing all rational numbers, the other all irrational numbers. Then the corresponding quotient space is the indiscrete topology over a set with two elements.

11. (a) Show that if for every two distinct elements $x, y \in X$ there is a continuous function $f : X \to \mathbb{R}$ such that $f(x) = 0$ and $f(y) = 1$ then the space X is Hausdorff.

 (b) Show if X is a T_1-space and if for every $x \in X$ and every closed subset F of X not containing x there is a continuous function $f : X \to \mathbb{R}$ such that $f(x) = 0$ and $f(F) = \{1\}$, then X is a T_3-space.

(c) Show that if X is a T_1-space and if for every two disjoint closed subsets F and G of X there is a continuous function $f : X \to \mathbb{R}$ such that $f(F) = \{0\}$ and $f(G) = \{1\}$ then X is a T_4-space.

[Note: the converses of (a) and (b) are not true. The converse of (c) is a major theorem – Urysohn's Lemma – and will be proven later.]

Solution. We prove (c); (a) and (b) require similar but easier arguments.
Assume the hypotheses of (c) are satisfied and choose any two disjoint subsets F and G that are closed in X. By assumption there is a continuous function $f : X \to \mathbb{R}$ such that $f(F) = \{0\}$ and $f(G) = \{1\}$. Then $U = f^{-1}(-\infty, 0.1)$ and $V = f^{-1}(0.9, \infty)$ are disjoint sets, open in X and separating F and G.

8.2 Regular Spaces and Normal Spaces

Solutions of the odd-numbered exercises.

---

1.   (a) Let X and Y be normal spaces. Show that $X \oplus Y$ is a normal space.
     (b) Let X and Y be completely regular spaces. Show that $X \oplus Y$ is a completely regular space.

Solution. (a) Let F and G be two disjoint sets, closed in $X \oplus Y$. Then $F \cap X$ and $G \cap X$ are disjoint and closed in X. Since X is normal, there are disjoint sets U_1 and V_1, open in X and such that $F \cap X \subset U_1$ and $G \cap X \subset V_1$. By symmetry, there are disjoint sets U_2 and V_2, open in Y and such that $F \cap Y \subset U_2$ and $G \cap Y \subset V_2$. Then the sets $U = U_1 \cup U_2$ and $V = V_1 \cup V_2$ are disjoint, open in $X \oplus Y$, and such that $F \subset U$, $G \subset V$. Hence $X \oplus Y$ is normal.
     (b) Let F be closed in $X \oplus Y$, and let z be a point out of F. We want to show that there is a continuous $f : X \oplus Y \to [0,1]$ such that $f(z) = 0$ and $f(F) = \{1\}$. The sets $F \cap X$ and $F \cap Y$ are closed in X and Y respectively. Since X and Y are completely regular, there are continuous functions $g : X \to [0,1]$ and $h : Y \to [0,1]$ such that $g(z) = 0 = h(z)$ and $g(F \cap X) = \{1\} = h(F \cap Y)$. Define $f : X \oplus Y \to [0,1]$ to be g over X and h over Y. Then f is continuous and $f(z) = 0$, $f(F) = \{1\}$

3. Show that every closed subspace of a normal space is normal.

Solution. Let X be normal and let Y be a closed subspace of X. Then every closed subset of Y is also closed in X. Hence if F and G are two closed, disjoint subsets of Y, they are also closed (and disjoint) in X. Since X is normal there are disjoint open subsets U and V in X such that $F \subset U$ and $G \subset V$. Then $U \cap Y$ and $V \cap Y$ are open (and disjoint) in Y, and $F \subset U \cap Y$, $G \subset V \cap Y$.

5. Show that if X and Y are completely regular spaces, then so is $X \times Y$.

Solution. Let X and Y be completely regular spaces. Assume F is a closed subset of $X \times Y$, and assume $(a,b) \notin F$. Denote by W the complement of F in $X \times Y$. Then there are open subsets U of X and V of Y such that $(a,b) \in U \times V \subset W$. Since $(U \times V)^c = (U^c \times Y) \cup (X \times V^c)$, it follows that $a \notin U^c$ and that $b \notin V^c$. Since X is completely regular, there is a function $f : X \to [0,1]$ such that $f(a) = 0$, and $f(U^c) = \{1\}$. Similarly, since Y is completely regular there is a function $g : Y \to [0,1]$

such that $g(b) = 0$, and $g(V^c) = \{1\}$. Define $h : X \times Y \to [0,1]$ by $h(x,y) = \max\{f(x), g(y)\}$. This function is continuous. Observe that $h(a,b) = 0$, and that $h\left[(U^c \times Y) \cup (X \times V^c)\right] = 1$. Consequently $h\left[(U \times V)^c\right] = 1$. Since $U \times V \subset W$, we have $W^c \subset (U \times V)^c$, and so $F \subset (U \times V)^c$. Hence $h(F) = 1$ and we have established what we wanted.

7. Show that every locally compact Hausdorff space is regular.

Solution. Let X be a locally compact, Hausdorff space, let F be a closed subset of X and let x be a point in $X \setminus F$. Since X is locally compact, there is an open neighborhood U of x such that $\overline{U}$ is compact. If $\overline{U} \cap F = \varnothing$ then U and $\overline{U}^c$ are two open sets that separate x and F. Suppose $\overline{U} \cap F \neq \varnothing$ and denote $\overline{U} \cap F = G$. Then G is a closed subset of the compact set $\overline{U}$. By Proposition 1, 7.2, G is compact. Since X is Hausdorff, for every $y \in G$ there are open neighborhoods V_y and U_x^y separating y and x. Since G is compact, the cover $\{U_y : y \in G\}$ possesses a finite subcover $\{U_{y_1}, U_{y_2}, \ldots, U_{y_n}\}$. Then $\bigcup_{i=1}^{n} U_{y_i}$ and $\bigcap_{i=1}^{n} U_x^{y_i}$ separate and x. Hence, $\overline{U}^c \cup \bigcup_{i=1}^{n} U_{y_i}$ and $U \cap \bigcap_{i=1}^{n} U_x^{y_i}$ separate F and x, and we proved that X is regular.

9. Let X be a normal separable space, and let E be a subset of X such that $|E| \geq 2^{\aleph_0}$. Assume in (a) and (b) that E has no accumulation points.

(a) Show that for every $Y \subset E$, Y and $E \setminus Y$ are disjoint and closed subsets of X.

(b) Let D be a countable dense subset of X. Define a mapping $f : \mathcal{P}(E) \to \mathcal{P}(D)$ as follows: for every $Y \subset E$ choose two disjoint open sets U_Y and $V_{E \setminus Y}$ such that $Y \subset U_Y$ and $E \setminus Y \subset V_{E \setminus Y}$, and set $f(Y) = U_Y \cap D$. Show that f is well defined and one-to-one.

(c) Deduce from (a) and (b) that every subset E, $|E| \geq 2^{\aleph_0}$, of a normal and separable space must have an accumulation point.

Solution.

(a) It is plain that the sets Y and $E \setminus Y$ are disjoint. Since E does not have any accumulation points, neither has any subset of it. On the other hand, it follows from Theorem 7 (b) in 3.2 that every set without accumulation points is closed.

(b) Since for every $Y \subset E$, Y and $E \setminus Y$ are disjoint and closed subsets of X, and since X is normal, the open sets U_Y and $V_{E \setminus Y}$ exist and so f is well defined.

To prove that f is one-to-one assume Y_1 and Y_2 are distinct subsets of E. By symmetry we may assume that there exist $y_1 \in Y_1$ such that $y_1 \notin Y_2$. The set $U_{Y_1} \cap V_{E \setminus Y_2}$ is open and, since y_1 is there, is not empty. Consequently $D \cap U_{Y_1} \cap V_{E \setminus Y_2} \neq \varnothing$. Hence $f(Y_1) = D \cap U_{Y_1}$ is also not empty. On the other hand, since $U_{Y_2} \cap V_{E \setminus Y_2} = \varnothing$, the set $f(Y_2) = D \cap U_{Y_2}$ is disjoint from $D \cap V_{E \setminus Y_2}$. Hence it is disjoint from $D \cap U_{Y_1} \cap V_{E \setminus Y_2}$. Thereby $f(Y_1) \neq f(Y_2)$, and so we proved that f is one-to-one.

(c) It follows from the definition of the linear order of the class of cardinal number (page 9) that $|\mathcal{P}(E)| \leq |\mathcal{P}(D)|$. Now $|\mathcal{P}(D)| = 2^{\aleph_0}$, hence $|\mathcal{P}(E)| \leq 2^{\aleph_0}$. On the other hand Proposition 3 in 1.2 gives that $|\mathcal{P}(E)| > |E|$, and so $|\mathcal{P}(E)| > 2^{\aleph_0}$. We have a contradiction.

Hence the set E must have accumulation points.

8.3 Normal Spaces and Subspaces

Solutions of the odd-numbered exercises.

---

1. Give an example of a compact space that is not normal.

Solution. Let X be any infinite set equipped with the co-finite topology. First we show that X is compact. Let $\mathcal{U}$ be an open cover of X and let U, $U \neq \varnothing$, be a member of that cover. Then U is missing finitely many elements $\{x_1, x_2, ..., x_n\}$ of X. Choose $U_i \in \mathcal{U}$ such that $x_i \in U_i$, $i = 1, 2, ..., n$. Then $\{U, U_1, U_2, ..., U_n\}$ is a finite subcover of $\mathcal{U}$. That X is not normal is obvious: any two non-empty open subsets of X avoid only finitely many elements of X, hence, since X is infinite, they cannot be disjoint. (In fact, the same argument implies that X is not a Hausdorff space.)

3. Show that ω is an accumulation point for TP, and that there is no sequence in $TP \setminus \{\omega\}$ converging to ω. [Hint for the second part: the argument is essentially the same as the one provided for the set Z in Example 2.]

Solution. Take an open neighborhood $(a, \omega]$ of ω. If $(a, \omega] \cap TP = \varnothing$, then a would be an element of $\mathbb{R}$ preceded (in the order of TP) by uncountably many elements. However, since $a < \omega$, this would contradict the choice of ω.

Suppose now that there is a sequence $\{b_n\}$ of elements in $TP \setminus \{\omega\}$ converging to ω. Then, as it is easy to see, $\bigcup_{i=1}^{\infty}[1, b_i] = [1, \omega]$. This is not possible since $\bigcup_{i=1}^{\infty}[1, b_i]$ is countable, while $[1, \omega]$ is not.

5. A space X is **perfectly normal** if it is normal and each closed subset is a countable intersection of open sets.
    (a) Show that being perfectly normal is hereditary.
    (b) Show that the Tychonoff plank is not perfectly normal.

Solution. Part (b) follows from (a) and Exercise 4. So, we prove (a): Take a subspace Y of X and a closed subset F of Y. Than $F = G \cap Y$ for some closed subset G of X. By assumption, $G = \bigcap_{i=1}^{\infty} U_i$ for some open subsets of X. Then $F = \left(\bigcap_{i=1}^{\infty} U_i\right) \cap Y = \bigcap_{i=1}^{\infty}(U_i \cap Y)$, and since the sets $U_i \cap Y$ are open in Y, we have achieved our goal.

Chapter 9: Urysohn, Tietze and Stone-Čech

9.1 Urysohn Lemma

Solutions of the odd-numbered exercises.

1. Show that a space X is normal if and only if every three closed pairwise disjoint sets can be separated with three open sets.

Solution. $\Leftarrow$ is obvious.
$\Rightarrow$ Suppose X is normal and let F, G and H be three pairwise disjoint closed sets. Then $F \cup G$ and H are disjoint closed sets. So, there are disjoint open sets U_1 and V containing $F \cup G$ and H respectively. Similarly, the normality of X implies that there are open disjoint sets U and W containing (as subsets) F and G respectively. Then $U_1 \cap U$, $U_1 \cap W$ and V are three open pairwise disjoint sets containing (as subsets) F, G and H respectively.

3. Show through a counterexample that we cannot replace the requirement that the sets A and B are closed in the statement of Urysohn's lemma by stipulating that these two sets are open.

Solution. Take $A = [0, 0.5)$ $B = (0.5, 1]$; then f as in the statement of the lemma cannot be extended to a map over $[0,1]$.

5. Prove that X is normal if and only if for every three pairwise disjoint closed sets A, B and C, and every three real numbers a, b and c, there is a continuous function $f : X \to \mathbb{R}$ such that $f(A) = \{a\}$, $f(B) = \{b\}$ and $f(C) = \{c\}$.

Solution.
Only $\Rightarrow$ deserves attention. Let us suppose that $a < b < c$. For $A \cup B$ and C^c there is an open set U such that $A \cup B \subset U \subset \overline{U} \subset C^c$. By Urysohn (and Exercise 4) there is $g : X \to \mathbb{R}$ such that $g(\overline{U}) = \{b\}$ and $g(C) = \{c\}$. Now consider $\overline{U}$: since it is closed, it is also normal. So, for A and B in it there is a set V that is open in $\overline{U}$ satisfying

$A \subset V \subset \overline{V} \subset B^c$. Find $h : X \to \mathbb{R}$ with $h(A) = \{a\}$ and $h(V^c) = \{b\}$ the complement

taken in $\overline{U}$). Define $f : X \to \mathbb{R}$ as follows: $f(x) = \begin{cases} h(x) \text{ if } x \in \overline{U} \\ g(x) \text{ if } x \in U^c \end{cases}$. Since it is obvious

that $\overline{U} \cap U^c$ is closed, and since h and g agree over this intersection, and since h and g are continuous, it follows by the gluing lemma that f is continuous. It is now easy to see that f maps A, B and C where we wanted.

7. Find an example of a normal space, countable many closed pairwise disjoint subsets A_i, $i \in \mathbb{N}$ in it, and countably many points a_i, $i \in \mathbb{N}$ in $\mathbb{R}$ such that there is no continuous mapping $f : X \to \mathbb{R}$ such that $f(A_i) = \{a_i\}$ for every $i \in \mathbb{N}$.

Solution. We can take $X = \mathbb{R}$, A_i-s to be points converging to some point, and a_i-s diverging points.

9.2 The Tietze Extension Theorem

Solutions of the odd-numbered exercises.

---

1. Let X be a normal space and let Y be any space homeomorphic to the interval $[0,1]$. Show that for every closed subset A of X, and every continuous function $f : A \to Y$ there is a continuous extension $\hat{f} : X \to Y$.

Solution. Let $h : Y \to [0,1]$ be a homeomorphism. Then $h \circ f : A \to [0,1]$ is continuous, and so, by Tietze, can be extended to $\widehat{h \circ f} : X \to [0,1]$. Consider the composition $h^{-1} \circ \left(\widehat{h \circ f} \right) : X \to Y$: for every a in A, we have $h^{-1} \circ \left(\widehat{h \circ f} \right)(a) = h^{-1}(h \circ f(a)) = f(a)$, and so it is the desired extension of f.

3. Deduce Urysohn's lemma from the Tietze extension theorem.

Solution. We pay attention to the non-trivial implication of Urysohn's lemma. Suppose A, B are two disjoint closed subsets of a normal space X. Then the mapping $f : A \cup B \to \mathbb{R}$ defined by $f(A) = \{0\}$, $f(B) = \{1\}$ is obviously continuous. By Tietze there is an extension $\hat{f} : X \to [0,1]$ and we are done.

5.   **(a)** Show that if X is a normal space, and if A is a closed subset of X, then every continuous mapping $f : A \to \mathbb{R}^n$ can be extend it to a continuous $\hat{f} : X \to \mathbb{R}^n$.
      (b) Generalize (a) as follows: Suppose Y is a space such that for every normal space X, every closed subset A of X, and every continuous mapping $f : A \to Y$, there is a continuous mapping $\hat{f} : X \to Y$ extending f. Show that the same is true for the product space Y^n in place of Y.

Solution. Extend each component of $g : A \to Y^n$. Then take the mapping that is made of these extensions as components.

7. Show that if X is a T_1 space such that for every closed subset A of X and every continuous mapping $f : A \to \mathbb{R}$ there is an extension $\hat{f} : X \to \mathbb{R}$ of f, then X is normal.

Solution. Let B and C be two disjoint closed subsets of X. Define $f : C \cup D \to \mathbb{R}$ by $f(C) = \{0\}$ and $f(D) = \{1\}$. It is then easy to check that f is continuous. Since $C \cup D$ is closed in X, it follows from the hypotheses of this exercise there is an extension $\hat{f} : X \to \mathbb{R}$. By Exercise 5 in 9.1 (which is essentially Urysohn lemma), X is normal.

9.3 Stone-Čech Compactification

Solutions of the odd-numbered exercises.

NOTE: in all of the exercises the spaces X, X_1, X_2 are completely regular T_1-spaces.

1. Show that if X is connected, then so is $\beta(X)$.

Solution. Since X is connected, so is $e(X)$. And closures of connected sets are connected!

3. Show that if $f : X \rightarrow [0,1]$ is a continuous function, then there is a continuous extension $F : \beta(X) \rightarrow [0,1]$.

Solution. This is almost a direct consequence of Theorem 3(c). The only part of the this exercise that deserves attention is the claim that the extension F has $[0,1]$ as its range too. This is a consequence of the proof of Theorem 3(c) – consider the diagram in Illustration 9.9 and observe that the range of each p_{f_0} in that diagram is the same as the range of f_0.

5. Let A and B be subsets of a space X, and let $f : X \rightarrow I$ be continuous function such that $f(A) = \{0\}$ and $f(B) = \{1\}$. Show that the closures $\overline{A}$ and $\overline{B}$ in $\beta(X)$ are disjoint.

Proof: By Exercise 3, f extends to $F : \beta(X) \rightarrow I$. Then, obviously $A \subset f^{-1}(\{0\})$, and $B \subset f^{-1}(\{1\})$, and since the sets on the right hand sided of $\subset$ are closed, we have that $\overline{A} \subset f^{-1}(\{0\})$ and $\overline{B} \subset f^{-1}(\{1\})$. That they are disjoint is now obvious.

7. Show that $\beta(\mathbb{N})$ is not metrizable.

Solution. In a metric space having and accumulation point implies existence of a sequence that converges to it. This is not so in $\beta(\mathbb{N})$ by Example 2.